C000030680

GRAIN BOUNDARY CONTROLLED
PROPERTIES OF FINE CERAMICS

JFCC Workshop Series:
Materials Processing and Design

Selected papers presented at the International Workshop on Fine Ceramics '92, 'Materials Processing and Design through Better Control of Grain Boundaries: Emphasizing Fine Ceramics', held in Nagoya, Japan, 12–13 March 1992.

GRAIN BOUNDARY CONTROLLED PROPERTIES OF FINE CERAMICS

JFCC Workshop Series:
Materials Processing and Design

Editors

KOZO ISHIZAKI

KOICHI NIIHARA

MITSUO ISOTANI

RENÉE G. FORD

ELSEVIER APPLIED SCIENCE
LONDON and NEW YORK

ELSEVIER SCIENCE PUBLISHERS LTD
Crown House, Linton Road, Barking, Essex IG11 8JU, England

WITH 26 TABLES AND 190 ILLUSTRATIONS

© 1992 ELSEVIER SCIENCE PUBLISHERS LTD

British Library Cataloguing in Publication Data

Materials Processing and Design: Grain
Boundary Controlled Properties of Fine
Ceramics.—(JFCC Workshop Series)
I. Ishizaki, Kozo   II. Series
620.1

ISBN 1-85166-952-3

Library of Congress CIP data applied for

No responsibility is assumed by the Publisher for any injury and/or damage to persons or property as a
matter of products liability, negligence or otherwise, or from any use or operation of any methods,
products, instructions or ideas contained in the material herein.

Special regulations for readers in the USA

This publication has been registered with the Copyright Clearance Center Inc. (CCC), Salem,
Massachusetts. Information can be obtained from the CCC about conditions under which photocopies
of parts of this publication may be made in the USA. All other copyright questions, including
photocopying outside the USA, should be referred to the publisher.

All rights reserved. No part of this publication may be reproduced, stored in a retrieval system, or
transmitted in any form or by any means, electronic, mechanical, photocopying, recording, or otherwise,
without the prior written permission of the publisher.

Printed in Great Britain by Galliard (Printers) Ltd., Great Yarmouth

Preface

This volume contains selected papers presented at a Workshop sponsored by the Japan Fine Ceramics Center on 'Materials Processing and Design through Better Control of Grain Boundaries: Emphasizing Fine Ceramics', held on 12–13 March 1992 in Nagoya, Japan. This Workshop brought together scientists and engineers from all over the world and from a variety of disciplines to focus on the application of grain boundary phenomena to materials processing and design. The topics covered include electronic materials, evaluation methods, structural materials and interfaces. Also included is an illuminating overview about the current status of work on grain boundary assisted materials processing and design, particularly for fine ceramics. The attendance of close on 100 participants from more than ten countries is indicative of the active interest in this topic.

This collection of high quality research papers presents current developments in the area of grain boundary related phenomena. These significant but little understood phenomena generated much lively discussion during the Workshop as well as fruitful interaction among those attending. Through the publication of this volume, the Japan Fine Ceramics Center hopes that many who were unable to attend this interesting and timely Workshop will benefit from its contents.

The Workshop was subsidized by the Japan Keirin Association through its Promotion Funds from KEIRIN RACE.

KOZO ISHIZAKI
KOICHI NIIHARA
MITSUO ISOTANI
RENÉE G. FORD

INTERNATIONAL WORKSHOP ON FINE CERAMICS '92

ORGANIZING COMMITTEE

Committee Chairman: Masaaki Ohashi

Vice-Chairmen: Kozo Ishizaki
Koichi Niihara
Mitsuo Isotani

Members: Isamu Fukuura
Kiyoshi Funatani
Hideyo Tabata
Noboru Yamamoto

Sumio Hirao
Jun-ichi Kon
Yukio Kubo
Hidetoshi Shibata
Jun-ichiro Tsubaki
Yoshimi Yamaguchi

Secretariat: Noriyuki Kosuge
Kaori Sakuma

Contents

Session III: Structural Ceramics I
(*Chairpersons:* K. NIIHARA and T. S. SUDARSHAN)

Session IV: Structural Ceramics II
(*Chairpersons:* R. RAJ, J. KRIEGESMANN and M. WATANABE)

Session V: Structural Ceramics III
(*Chairpersons:* I-W. CHEN and F. WAKAI)

Session VI: Interfaces
(*Chairpersons:* K. KENDALL and M. IWATA)

Session I

Electronic Ceramics

PROCESSING KINETICS AND GRAIN BOUNDARY ELECTRICAL PROPERTIES OF ELECTROCERAMICS VIA IMPEDANCE SPECTROSCOPY

E. A. COOPER, K. S. KIRKPATRICK, B. J. CHRISTENSEN, B. -S. HONG, and T. O. MASON
Northwestern University
Department of Materials Science and Engineering and Materials Research Center
2225 Sheridan Rd.
Evanston, IL 60208

ABSTRACT

In situ impedance spectroscopy (IS) is a powerful in situ monitor of processing kinetics for electroceramics. Examples of "parallel" interconnectivities encountered during processing include chemical bonding, sintering, solid state reactions, and decomposition reactions. "Series" interconnectivities occur during grain boundary processing, e.g. doping, oxidation/reduction, etc. Data from the processing of oxide superconductors illustrate the utility of IS for in situ processing studies of electroceramics.

INTRODUCTION

Impedance spectroscopy (IS) is an already robust field of research. The classic source text for IS is the book by Macdonald [1]. IS is used extensively in the fields of electrochemistry, solid electrolytes, and corrosion. In electroceramics, IS is a well known means of resolving grain boundary vs. bulk conduction in polycrystalline specimens. Little, if any, work has been reported on *in situ* IS during the processing of electroceramics. Furthermore, the bulk of IS research has involved ionic rather than electronic conductors. The present study presents an overview of our recent efforts in the area of in situ processing of mainly electronic ceramics by IS.

In order to adequately address the underlying factors influencing inhomogeneous conduction in polyphase and/or polycrystalline materials, it is necessary to introduce the concepts of interconnectivity patterns (IPs), mixing laws, and equivalent circuits. The recent review/tutorial by McLachlan et al. [2] is extremely helpful in this regard.

Two phases, one a conductor and one an insulator, can be put together in a variety of interconnectivity patterns (IPs). If we represent the two interconnectivities as h-l, where h=high conductivity and l=low conductivity, there are 16 possible combinations. (McLachlan et al. [2] catalogue only 10 IPs, but make no distinction between *h-1* and *1-h* IPs.) For example, a 1-3 IP would have one-dimensional conductor rods imbedded in a 3-D insulator matrix. A 3-0 IP represents isolated insulator particles embedded in a 3-D conductor matrix. Since all of the materials we will consider are macroscopically isotropic, any IP with 1 or 2 is disallowed, leaving 0-0, 0-3, 3-0, and 3-3 combinations only.

McLachlan et al. [2] also catalogued two phase mixing laws for conductivity. For example, Figure 1 displays "series" and "parallel" mixing laws with a 7 order of magnitude difference in conductivity between the two phases. We will continue to employ the terms "series" and "parallel" to describe the limiting cases in the present study. Rigorously, however, series and parallel laws are for anisotropic IPs (e.g. 2-2), and different limits, the so-called Hashin-Shtrikman (HS) upper and lower bounds have been determined for isotropic IPs [3], as shown. These are quite similar to the series and parallel curves (see Figure 1) which are easier to conceptualize.

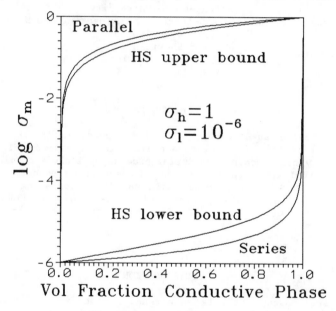

Figure 1. Conductivity of a binary composite as a function of the volume fraction conductive phase for different interconnectivity patterns.

Finally, the frequency response of "series" and "parallel" situations must be considered. This can be done with the aid of "equivalent circuits," resistor-capacitor (RC) combinations capable of simulating the behavior. In Figure 2 are shown equivalent circuits and impedance diagrams for parallel (e.g. 3-3) and series (e.g. 0-0) IPs. We believe the 3-0 and 0-3 IPs will exhibit parallel and series responses, respectively, if the insulator volume fraction is small in the 0-3 case. The latter is the classic "brick-layer" model where resistive grain boundaries surround the more conductive grain interiors. The equations on Figure 2 can be used to obtain useful information about the properties and even the volume fractions of the two phases, as demonstrated below.

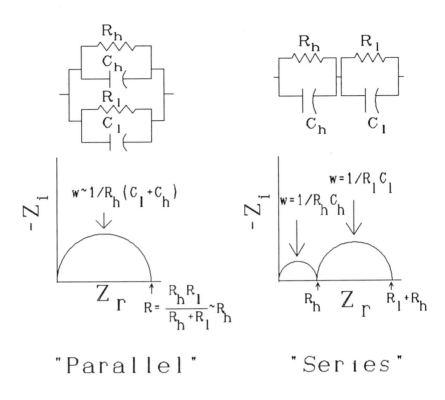

"Parallel"   "Series"

Figure 2. Electrical circuit analogues of interconnectivity patterns (top) and their resultant impedance plots (bottom). w is frequency in radians.

EXPERIMENTAL

Because sample preparation differs from system to system, experimental details will be given for each case study below. Reference will also be made to already published work, where experimental details can be found in the appropriate literature (see Table 1). The experimental apparatus is described elsewhere [4]. It consists of a high temperature specimen holder in which pressed pellets are held under slight compression between silver or platinum foils. The entire assembly is constructed of fused quartz to enable rapid insertion into a preheated tube furnace under flowing atmosphere. Leads from the contact foils were connected to an impedance analyzer (Hewlett-Packard Co., Palo Alto, CA, model 4192A) computer controlled by an 80386-based system. A unique feature of the data acquisition system was the ability to perform an abbreviated frequency scan (20 points) covering 11 MHz to 100 Hz in less than 15 seconds. This enables IS to monitor transient phenomena during procesing in "real time." When rapid scans were not needed, complete 13 MHz to 5 Hz scans were performed. Impedance data (frequency, real, imaginary) were imported to "Equivalent Circuit," a dedicated software program for RCL analysis [5].

CASE STUDIES

"Parallel" Interconnectivities

The overall conductivity (σ) of composites with "parallel" interconnectedness can be described with the equation:

$$\sigma - \sum_i \sigma_i \, \phi_i \, \beta_i \qquad (1)$$

where, for each phase i, σ is conductivity, ϕ is volume fraction, and β is an interconnectivity parameter between 0 and 1. This equation is derived from porous rock and cement/concrete literature (see [6] for a discussion). When dominated by the high conductivity phase (h), this equation reduces to:

$$\sigma \approx \sigma_h \, \phi_h \, \beta_h \qquad (2)$$

We have made extensive studies of the underlying parameters during processing reactions of electroceramics, as listed in Table I.

TABLE I
Processing Reactions Exhibiting "Parallel" Interconnectivities

PROCESS	σ_h	ϕ_h	β_h	Example
Chemical Bonding	=f(t)	=g(t)	=h(t)	Ref. 8
Sintering	typically constant	gradual increase, $0.5 < \phi_h < 1.0$	strong dependence on neck size	Refs. 9,10
Decomposition Reactions	typically constant	approx. $\sigma(t)/\sigma(o)$	approx. unity	Ref. 11
Solid State Reactions	typically constant	approx. $\sigma(t)/\sigma(\infty)$	approx. unity	This Work

The first instance is that of chemically bonded ceramics, i.e. cement-based materials. (Though not usually considered "electroceramics," cement-based materials are under consideration for electronic packaging [7].) All three parameters in Eq. 2 can change with time during the chemical bonding process. This is an ionic conductor system where the conductivity, phase fraction, and interconnectivity of the conductive pore fluid change with time. Fortunately, alternate techniques exist to determine how σ_h and ϕ_h change with time in order to establish the time dependence of β_h, the microstructural parameter. Further details of the analysis are given elsewhere [8]. Discounting a low frequency arc associated with electrode effects, characteristic single-arc "parallel" behavior is observed in IS (see Figure 2). This approach may be applicable to other chemically bonded ceramic systems.

In sintering, σ_h is usually constant, and ϕ_h changes within a narrow and predictable range, i.e. 0.5 to 1, as densification proceeds. What changes dramatically with time is the interconnectedness, β_h. This changes as coordination number and neck size (X) change during sintering. Witt has demonstrated an approximately $\sigma \propto X$ dependence in the sintering of monosized oxide particles [9]. We have reported a study of sintering kinetics in Bi_2O_3-doped ZnO via IS. Again, characteristic single-arc, "parallel" behavior was observed. The rate of increase of conductivity with time at a given composition increased dramatically upon crossing the liquidus or eutectic temperature, signaling a change from solid state to liquid state sintering. From the liquidus temperatures, a proposed phase diagram for the ZnO end of the ZnO-Bi_2O_3 system was developed. Further details are given in [10].

Characteristic single-arc, "parallel" IS curves were obtained during the decomposition of $YBa_2Cu_3O_7$ in CO_2, an extremely important processing consideration. Of the parameters in Equation 2, σ_h is expected to remain essentially constant, β_h is believed to be approximately unity, and ϕ_h decreases from unity to zero as decomposition proceeds. By parallel quantitative X-ray diffraction, it was shown that the normalized conductivity, $\sigma(t)/\sigma(o)$, scaled with the volume fraction of undecomposed superconductor. The most important outcome of this work is that decomposition is at pore surfaces and/or select grain boundaries rather than at all grain boundaries. Otherwise, the double-arc, "series" behavior of Figure 2 would have been observed. Further details are given in [11].

The final example, that of solid state formation reaction of superconducting oxides from various precursors, is an excellent illustration of IS to study processing kinetics. Powders of La_2O_3 (Aldrich, Milwaukee, WI, 99% pure) and CuO (Aesar, Johnson Matthey, Ward Hill, MA, 99.99% pure) were ball milled in cyclohexane to eliminate $La(OH)_3$ formation. Preparations were conducted in a glove box to prevent moisture-sensitivity. Reaction pellets were uniaxially pressed in a 7.5 mm die at 125 MPa. Pellets were divided into 2-3 pieces, one of which was used for quantitative X-ray diffraction using appropriate standards. The other pieces were subjected to in situ IS upon insertion into a preheated furnace, as described above. A typical series of IS curves as a function of time is presented in Figure 3. A single arc with a capacitance in the 100 pF range was consistently obtained, indicative of bulk vis-a-vis interfacial response. As can be seen, the arc rapidly shrinks with time. The low frequency (DC) intercept (normalized to infinite time) is plotted vs. fraction of product in Figure 4. The appearance is strikingly similar to the "parallel" or HS-upper bound behavior of Figure 1, confirming the 3-3 or 3-0 interconnectivity. A more rigorous treatment of the appropriate mixing law behavior, taking into account the conductivities of the reactants (CuO is also relatively conductive) is given in [11].

In summary, "parallel" interconnectivity situations in processing can be studied by IS if there is a sufficient difference in conductivity between the high and low conductivity phases. A single bulk arc will be observed. The low frequency (DC) intercept provides useful information concerning the volume fraction of the conductive phase (when σ_h and β_h are relatively constant) as in the case of chemical bonding, formation, and decomposition reactions, or about the interconnectedness of the conductive phase (when σ_h and ϕ_h are relatively constant) as in the early stages of sintering. In this case, β_h provides insight into the sintering processes.

"Series" Interconnectivities

IS is now routinely employed to study the grain boundary electrical behavior in electroceramics. For example, soft magnetic ferrites undergo an oxidative annealing step to increase overall resistivity. Cheng [12] demonstrated that post-anneal IS spectra exhibit distinct grain boundary arcs which grow in size with the length of annealing and the magnitude of overall resistivity. Similarly, ZnO-based varistors [13] and $BaTiO_3$-based PTCR

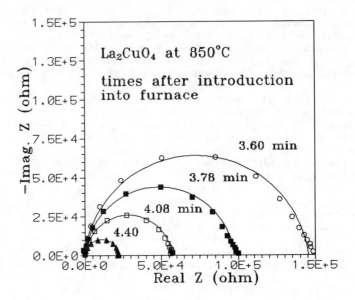

Figure 3. Impedance curves showing the decrease in bulk arc size with time
and reaction.

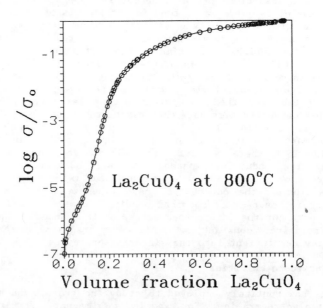

Figure 4. Normalized conductivity versus volume fraction of conductive phase
formed during reaction.

thermistors [14] exhibit distinct grain boundary arcs. In the latter case, it was shown that the grain boundary arc grows dramatically as temperature crosses T_c, confirming that the PTCR effect is associated with property changes at the grain boundaries [14].

All of these studies were post-processing, however. Only Schlouler et al. [15] attempted to employ IS as an *in situ* processing monitor for yttria-stabilized ceramics. These authors were able to observe dissolution of the dopant into the host, elimination of porosity, and segregation of impurities to the grain boundaries. We believe IS can be a powerful *in situ* monitor of grain boundary engineering in electroceramics.

The data of Figure 5 illustrate the capability for IS to monitor grain boundary changes during processing. What are being presented are IS curves for the reaction of La_2O_3 and CuO, as outlined above, but at a lower temperature (750°C) and at early times. Separate grain boundary and grain interior arcs are discernable. Both are shrinking with increasing time, but moreso the grain boundary arc. If we assume identical dielectric constants for grain boundaries (gb) and grain interiors (gi), we can arrive at an estimate for ratio of grain boundary thickness to average grain size [16],

$$\frac{C_{gi}}{C_{gb}} \approx \frac{d}{D} \qquad (3)$$

where D is average grain size. This ratio is plotted vs. time in Figure 6.

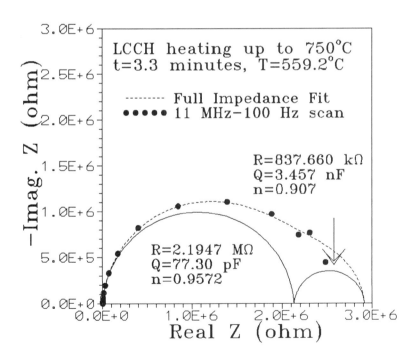

Figure 5a. Impedance plot at 0.055 hrs showing the response of grains and grain boundaries.

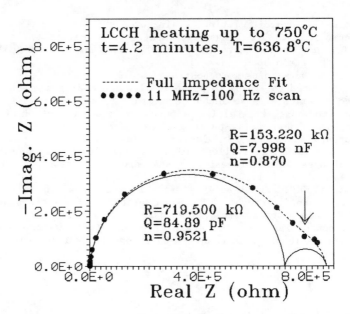

Figure 5b. Impedance plot at 0.070 hrs showing the response of grains and grain boundaries.

What this illustrates is that the grain boundary "phase" (this could be the same phase, but with a different composition or defect content) disappears rapidly with time. From the same "brick layer" model [16], the resistivity ratio can be obtained,

$$\frac{\rho_{gb}}{\rho_{gi}} \approx \frac{R_{gb}}{R_{gi}} \frac{D}{d} \qquad \qquad (4)$$

The grain boundary "phase" is 20 to 100 times as resistive as the grain interiors, depending upon time.

In summary, IS can resolve grain boundary property variations during high temperature processing. Since the volume fraction of the grain boundary phase is miniscule, overall conductivity should fall on the "series" limit in Figure 1. This explains why a very small fraction of grain boundary phase can produce such a marked decrease in conductivity (or increase in resistivity). Although the example cited involved transient grain boundary behavior, we believe that IS will be increasingly employed as an in situ probe of stable grain boundary electrical modifications. Ongoing research on ZnO-based varistor and thermistor formulations will be reported separately.

CONCLUSIONS

The frequency response of electroceramic "composites" can be described within the "interconnectivity pattern" framework using standard mixing laws and equivalent circuit concepts. Impedance spectroscopy is a powerful tool

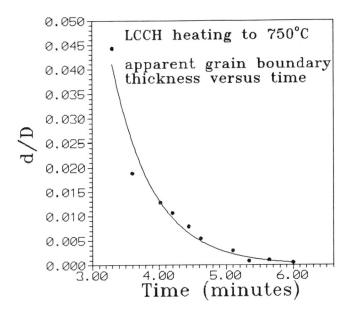

Figure 6. Ratio of the apparent grain boundary size (d) to that of the grain interior (D) as a function of time from the capacitance ratios.

for monitoring in situ changes in the amounts and distribution of phases during processing. Two limiting cases are encountered--"parallel" interconnectivities, where single-arc impedance spectra are obtained and conductivity scales with the volume fraction (ϕ_h) of the conducting phase (σ_h and β_h constant), as during chemical bonding, formation reactions, and decomposition reactions, or conductivity scales with the interconnectivity (β_h) of the network (σ_h and ϕ_h constant), as during the initial phase of sintering; and "series" interconnectivities, where dual-arc impedance spectra are obtained, each arc providing information concerning the individual components, e.g. grain boundary vs. bulk. In certain cases, such as the solid state reaction described above, both "parallel" and "series" aspects occur simultaneously. IS has important technological ramifications for monitoring the processing reactions in electroceramics, including consolidation processes and grain boundary engineering.

ACKNOWLEDGMENTS

The superconductor IS work was supported by the National Science Foundation under contract No. DMR-8809854 (EAC, TOM). The authors also acknowledge the support of the NSF under contract DMR-8808432 (BJC, TOM) and DMR-8821571 (BSH, TOM), and the U.S. Dept. of Energy under contract W-31-109-Eng-38 and the Division of Educational Programs, Argonne National Lab (KSK).

REFERENCES

1. Macdonald, J. R., <u>Impedance Spectroscopy: Emphasizing Solid Materials and</u>

Systems, Wiley, New York, 1987.

2. McLachlan, D.S., Blaszkiewicz, M., and Newnham, R.E., Electrical Resistivity of Composites, J. Am. Ceram. Soc., 1990, 73, 2187-2203.

3. Hashin, Z. and Shtrikman, S., A Variational Approach to the Theory of the Effective Magnetic Permeability of Multiphase Materials, J. Appl. Phys., 1962, 33, 3125-31.

4. Cooper, E. A., Exploration of Selected Cuprate Superconductor Reaction Kinetics Using XRD and Impedance Spectroscopy, Ph.D. dissertation, Northwestern University, Evanston, IL, June, 1992.

5. Boukamp, B. A., EQUIVALENT CIRCUIT (EQUIVCRT.PAS), Dept. of Chemical Tech., Univ. of Twente, P.O. Box 217, 7500 AE, Enschede, Netherlands, 1988.

6. Garbozci, E. J., Permeability, Diffusivity, and Microstructural Parameters: A Critical Review, Cem. Conc. Res., 1990, 20, 591-601.

7. Leigh, D. W., Payne, D. A., and Young, J. F., Preparation and Properties of Hardened Cement Materials for Electrical Applications, Adv. in Ceramics, 1989, 26, 255-63.

8. Christensen, B. J., Mason, T. O., Jennings, H. M., Bentz, D. P., and Garbozci, E. J., Experimental and Computer Simulation Results for the Electrical Conductivity of Portland Cement Paste, Proc. Mat. Res. Soc. Symp., in press.

9. Witt, C. A., The Sintering Behavior of Magnetite, Ph.D. dissertation, Northwestern University, Evanston, IL, June, 1987.

10. Kirkpatrick, K. S., Mason, T. O., Balachandran, U., and Poeppel, R. B., Impedance Spectroscopy Study of Sintering in Bi-Doped ZnO, J. Am. Ceram. Soc., submitted.

11. Cooper, E. A., Gangopadhyay, A. K., Mason, T. O., and Balachandran, U., CO_2 Decomposition Kinetics of $YBa_2Cu_3O_{7-x}$ via In Situ Electrical Conductivity Measurements, J. Mater. Res., 1991, 6, 1393-7.

12. Cheng, H. F., Modeling of Electrical Response for Semiconducting Ferrite, J. Appl. Phys., 1984, 56, 1831-7.

13. Alim, M. A., Admittance-Frequency Repsonse in Zinc Oxide Varistor Ceramics, J. Am. Ceram. Soc., 1989, 72, 28-32.

14. Tseng, T. and Wang, S. H., The A.C. Electrical Properties of High-Curie-Point Barium Lead Titanate PTCR Ceramics, Mater. Lett., 1990, 9, 164-168.

15. Schouler, E. J. L., Mesbahi, N., and Vitter, G., In Situ Study of the Sintering Process of Yttria-Stabilized Zirconia by Impedance Spectroscopy, Solid State Ionics, 1983, 9/10, 989-96.

16. Bonanos, N., Steele, B. C. H., Butler, E. P., Johnson, W. B., Worrell, W. L., Macdonald, D. D. and McKubre, M. D. H., Applications of Impedance Spectroscopy, Ch. 4 in Impedance Spectroscopy: Emphasizing Solid Materials and Systems, J. R. Macdonald, Ed., Wiley, New York, 1987, 198-205.

DISCONTINUOUS GRAIN GROWTH AND THE GRAIN BOUNDARY IN A SOLID-SOLID REACTION GROWTH METHOD FOR FERRITE SINGLE CRYSTALS

Minoru Imaeda, Yoshinari Kozuka and Soichiro Matsuzawa
Electronics & Optoelectronics Research Laboratory
NGK Insulators, Ltd.
2-56 Suda-cho, Mizuho-ku, Nagoya, 467 Japan

ABSTRACT

A method for growing single crystals by a solid-solid reaction has been developed for Mn-Zn ferrite and YIG (yttrium iron garnet). This method consists of joining a polycrystal with a single crystal seed and heating the joined body. Homogeneous single crystals are obtained readily. In this method, it is very important to control grain boundary phenomena such as discontinuous grain growth, grain size, pores, and impurities. In this paper, the relationships between grain growth and starting materials, impurities in Mn-Zn ferrite, and the stoichiometry of YIG are discussed. The principles of this solid-solid reaction growth method are described based on experimental data.

INTRODUCTION

We have developed a new method to grow single crystals in the solid phase, called solid-solid reaction. We are now commercially producing Mn-Zn ferrite single crystals for magnetic recording heads by this method. It is also applicable to yttrium iron garnet (YIG) single crystals, which are used as optical isolators and optical magnetic field sensors.

This method is based on joining the polycrystal with a single crystal seed and heating the joined body. A single crystal is then obtained via solid state processing. The key characteristics of this solid-solid reaction compared with single crystal growth methods via the melt are 1) homogeneous composition and properties, 2) high purity, and 3) low cost.

It is most important in this solid-solid reaction method to control grain boundary phenomena, such as discontinuous grain growth, grain size, pores, and impurities. We have had some data on the single crystal growth of these materials, but we haven't clearly understood the principle. In this paper, the production method and the phenomena related to the grain boundary are described, followed by a discussion of the principle of the solid-solid reaction.

PRODUCTION METHOD

The production method for Mn-Zn ferrite differs somewhat from that of YIG,
but the main process is similar. Therefore, the process of producing a Mn-
Zn ferrite single crystal is described first.

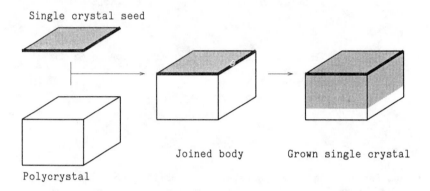

Figure 1. The process of single crystal growth by solid-solid reaction

Mn-Zn ferrite single crystal

Iron oxide (Fe_2O_3), manganese oxide (MnO), and zinc oxide (ZnO) with a
purity of $> 99.99\%$ are mixed to make the desired composition, for example
53 mol% Fe_2O_3, 31 mol% MnO, and 16 mol% ZnO. The mixture is calcined at 1200
℃ for 2 hours and milled in an iron container with iron balls. Next the
powder is molded and sintered at 1320℃ for 4 hours under equilibrium
oxygen partial pressure. The resulting polycrystal has an average grain
size of about 10 μm and is more than 99.98% dense.

   The ferrite polycrystal thus obtained is cut into a 5x25x10 mm shape.
The single crystal seed (5x25x1 mm) and the polycrystal are polished by
diamond abrasive powder on a tin disc to a roughness of $< 0.05 \mu$m. Both of
the polished surfaces are joined by using a 6N HNO_3 solution as adhesive
(see Figure 1), and the joined body is heated at 1370 ℃ for 8 hours under
equilibrium oxygen partial pressure.

   During the heating period, the boundary of the single crystal and the
polycrystal migrates and the joined body becomes a large single crystal.
Thousands of single crystals can be produced in one furnace in a short time
compared with conventional methods, such as the Bridgeman method.

YIG single crystal

There are two different parts to the method for growing a YIG crystal. First,
the coprecipitated powder is used as the starting material. By mixing
yttrium oxide and iron oxide powder, it is impossible to obtain a more than
99.0% dense polycrystal. In the coprecipitated powder process, solutions of
iron(II) sulfate and yttrium nitrate are prepared in the stoichiometry of

$Y_3Fe_5O_{12}$ and mixed with a solution of ammonia in water. The coprecipitated hydroxide powder obtained is calcined at 1200 °C for 2 hours. It can be sintered to more than 99.9% density at 1400 °C for 8 hours.

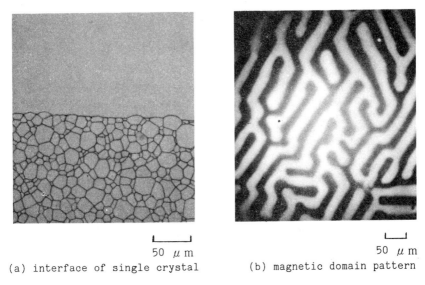

| | |
| (a) interface of single crystal | (b) magnetic domain pattern |

50 μ m   50 μ m

Figure 2. A single crystal of YIG grown by solid-solid reaction

Next, the solid-solid reaction is carried out in a HIP furnace. The stoichiometric joined body is heated at 1500 °C and 1500 atm in an argon atmosphere for 8 hours. To use it as an optical material, the porosity of the crystal must be reduced. HIP treatment during the growing process produces poreless and transparent YIG single crystals, with 0.002% porosity. The interface of the grown single crystal and the polycrystal, in which the growth of the crystal is interrupted, is shown in Figure 2(a). It was observed by a transparent polarized infra-red microscope. Figure 2(b) shows the large magnetic domain patterns plus the presence of a few pores. The large magnetic domain patterns, with a width of 200 - 400 μ m, indicate that the YIG single crystal grown by the solid-solid reaction is only slightly distorted.

RESULT AND DISCUSSION

In the development of this single crystal growth method we discovered certain phenomena associated with grain boundary behavior, discussed below.

Grain growth and starting materials of Mn-Zn ferrite

In the development of Mn-Zn ferrite single crystal growth by the solid-solid reaction process, we found that there are two type of grain growth phenomena, as shown in Figure 3. One is normal grain growth(A), in which a single crystal can not be grown, and the other is discontinuous grain growth(B), which is the basic mechanism for single crystal growth.

To grow single crystals by this solid-solid reaction, it is necessary that the grain size of the polycrystal is retained during crystal growth, because nuclei growth temperatures increase as the grain size increases. By using type(B) polycrystals, the average grain size is kept to about 10 μm below 1390℃, which is the nucleation temperature of discontinuous grain growth. The dotted line(C) represents the extent of single crystal growth of a type(B) polycrystal after 30 min treatment in the furnace. The nuclei growth temperature is about 1360℃, therefor single crystals can be grown between 1370 and 1380 ℃ as shown in Figure 3.

The difference between the two polycrystals is the starting iron oxide material. Iron oxide with a spinel structure, such as magnetite and hematite made from magnetite by calcination, is used as the starting material for a type(B) polycrystal. But if hematite made from iron sulfate by calcination is used, the sintered polycrystals behave like type(A).

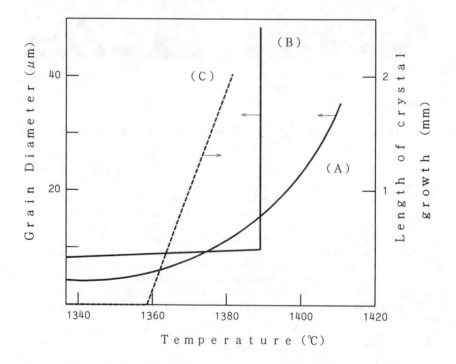

Figure 3. The difference between grain growth behavior and nuclei growth in Mn-Zn ferrite

Grain growth and impurities in Mn-Zn ferrite

The growing method is so sensitive to impurities that the total amount of impurities must be reduced to under 30 ppm when growing Mn-Zn ferrite single crystals. Figure 4 shows the relationship between the SiO_2 impurity, the nucleation temperature of discontinuous grain growth, and the porosity of the single crystal.

When the impurities are below 100 ppm(A), the nucleation temperature increases, therefore the growth temperature must be raised and more pores remain in the crystal.

When the impurities are above 300 ppm(B), the nucleation temperature decreases to below 1300 ℃. This is not considered to be growth in the solid phase. A small amount of liquid phase is seen in the grain boundary. Previously encountered discontinuous grain growth is considered to be of this type. It will not be used for single crystal growth, because the porosity is large and thus it is less homogeneous.

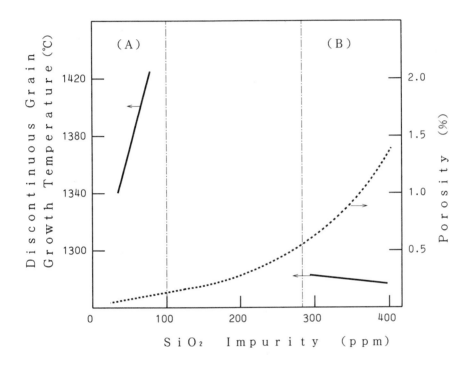

Figure 4. The difference in grain growth behavior and porosity due to
impurities in the Mn-Zn ferrite

Grain growth and the stoichiometry of YIG

YIG characteristically has an incongruent melting point and a very narrow stoichiometric region compared with Mn-Zn ferrite, a perfect solid solution. So the composition and growing temperature must be controlled precisely. The stoichiometry of YIG must be controlled within 0.05 mol%.

When more yttrium is present than the stoichiometric quantity, a second phase of orthoferrite is observed at the grain boundary. Then discontinuous grain growth doesn't occur at all, and thus it is impossible to grow single crystals.

In contrast, the iron rich sample contains a second phase of hematite as a liquid phase at the grain boundary. In this case, discontinuous grain growth takes place as with a stoichiometric sample. But the interface of the single crystal and the polycrystal form a straight line, which is very different from the solid phase reaction, and the grown crystal includes a second phase. The mechanism of single crystal growth is considered to be the same as for the Mn-Zn ferrite with SiO_2 impurity.

CONCLUSION

We have developed two types of ferrite single crystals by a solid-solid reaction method, and have described some of the phenomena related to the grain boundary. The principle of this solid-solid reaction is not understood yet, but we now believe the key points are the following:
 (1) A polycrystal that shows discontinuous grain growth must be used, and its grain growth behavior depends on the starting material used for Mn-Zn ferrite.
 (2) The joined single crystal seed and the polycrystal must be treated above the nuclei growth temperature and below the nucleation temperature for discontinuous grain growth.
 (3) The presence of impurities, which produce a second phase, must be reduced to as small an amount as possible.

REFERENCES

(1) Matsuzawa, S. and Kozuka, Y., Method for producing ferrite single crystals by solid-solid reaction. Adv. Ceram., 1985, 15, 527-532.
(2) Imaeda, M. and Matsuzawa, S., Growth of yttrium iron garnet single crystal by solid-solid reaction. Proc. 1st Japan SAMPE Symposium., 1989, 419-424.

THERMAL PROPERTIES OF AlN SINTERED AT LOWER TEMPERATURE WITH ADDITIVES

N.ICHINOSE and I.HAZEYAMA
School of Science and Engineering, Waseda University
3-4-1 Ohkubo, Sinjuku-ku, Tokyo

ABSTRACT

It is well known that the sintering temperature of aluminum nitride (AlN) ceramics is above 1800℃. Therefore, it would be desirable to sinter AlN ceramics at lower temperature for energy conservation. Sintering additives, such as CaO, CaF_2, Y_2O_3, and YF_3, are essential components for the densification of AlN and to obtain higher thermal conductivity. Y_2O_3 and YF_3 were investigated as sintering additives at 1600 ℃ and 1800 ℃. The additives YF_3 was found to trap oxygen at 1600 ℃.

INTRODUCTION

As sintering additives for AlN, various materials were studied [1]. CaO, CaF_2, Y_2O_3, YF_3, and AlF_3 were investigated as potential additives for lowering the sintering temperature [2]. It was found that CaO and CaF_2 were effective for lowering the sintering temperature. AlF_3 was found to be more effective if it was added with the other additives. Furthermore, the difference between CaO and CaF_2 addition was investigated, particularly with respect to its effect in trapping oxygen [3]. Oxygen was present on the surface of the AlN lattice substituting for nitrogen. In this process, vacancies occur on the aluminum sites. The lattice parameter of AlN decreases as the oxygen content increases [4]. Therefore, it is possible to estimate the solute oxygen value by measuring the lattice parameter. To obtain higher thermally conductive AlN, the amount of oxygen present in the AlN lattice must be controlled. In this paper, the effect of Y_2O_3 and YF_3 addition on the thermal conductivity of AlN was investigated, using samples sintered at 1600 ℃ and 1800 ℃.

EXPERIMENTAL

Sample preparation
The samples were fabricated by means of a conventional procedure as follows; The starting materials are AlN (Tokuyama Soda Co.,Ltd., F grade; oxygen content $\leqq 1.0$ %, average grain size 1.8 μm), Y_2O_3, YF_3 (Soekawa Chemical Co.,Ltd., purity 3N), and AlF_3 powder (High Purity Chemical Co.,Ltd., purity 3N). After weighing (Table 1), these powders were mixed by ball milling in ethanol using nylon balls,

for 4 hours. The slurry was dried and on forming a slurry, acrylic binder was added with 1-1-1 trichloroethane. After drying, the powder mixture was pressed into 12 mm diameter discs, at about 500 kg / cm ^{2}.

Table 1. Additive content in AlN

Sample	Additives & Content
1	Y_2O_3 0.5, 1.0, 1.5, 3.0 wt %
2	YF_3 0.5, 1.0, 1.5, 3.0 wt %
3	Y_2O_3 1.0 wt % AlF_3 1.0 wt %
4	Y_2O_3 3.0 wt % AlF_3 1.0 wt %
5	YF_3 1.0 wt % AlF_3 1.0 wt %
6	YF_3 1.0 wt % AlF_3 1.0 wt %

The samples were heated at 700℃ in flowing N_2 gas in order to burn out the binder. After this treatment, the samples were sinterd at 1600℃ and 1800℃ for 2, 4, 8, 16, and 32 hours in a N_2 atmosphere.

Evaluation
The density was calculated by Archimedes' principle after weighing in water and air. Thermal conductivities were measured by laser flash using a ruby laser (λ=0.6943 mm) and a liquid-nitrogen-cooled InSb infrared detector at room temperature. X-ray powder diffraction was used to identify the crystallographic phases after crushing the sinterd body in an agate mortar. Lattice constants of AlN were measured using metallic silicon as the internal standard. Differential Thermal Analysis and Thermogravimetry (DTA-TG) were used to investigate the decomposition of the fluoride.

RESULTS AND DISCUSSION

The densities of the samples sinterd at 1600℃ for 2, 4, 8, and 16 hours were similar, but all did not reach a theoretical density. To densify AlN by sintering at this temperature with Y_2O_3, YF_3, and AlF_3 additives, requires longer sintering times.

Thermal diffusivities of these samples increased with increasing sintering time. The diffusivities of samples with Y_2O_3 (1.0 wt%), Y_2O_3 (1.0 wt%)-AlF_3 (1.0 wt%), YF_3 (1.0 wt%), and YF_3 (1.0 wt %)-AlF_3 (1.0 wt %) additives sinterd for 2 hours were 15, 17, 21, and 20 cm²/sec, respectively. And therefore samples sinterd for 4 hours were 18, 21, 26, and 26 cm ^{2}/sec, respectively.

Figure 1 shows the c-axis lattice parameter of the samples with Y_2O_3 (1.0 wt%)-AlF_3 (1.0 wt%) and YF_3 (1.0 wt%) -AlF_3 (1.0 wt%) additives. In the case of the YF_3-AlF_3 additives, it increases with increasing sintering time in the range of 0-2 hours, and saturates above 2 hours. On the other hand, for the sample with Y_2O_3-AlF_3, it increases markedly in the range of 4-8 hours. At 2 and 4 hours, the lattice constant becomes smaller than that of the raw AlN powder. This could be due to the presence of dissolved oxygen in the AlN.

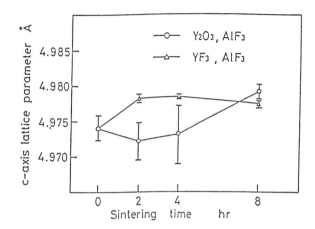

Figure 1. The c-axis lattice parameter of AlN sintered at 1600 °C.

Figure 2 shows the c-axis lattice parameter of AlN sintered at 1800°C. From this figure it can be seen that there is no difference in the effect of adding Y_2O_3-AlF_3 or YF_3-AlF_3. It is known that the Y-Al-O phase promotes liquid phase sintering at this tempeture, therefore the samples become densified.

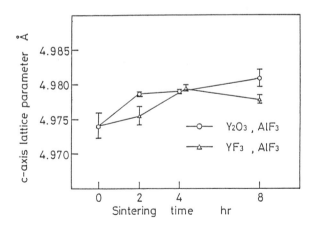

Figure 2. The c-axis lattice parameter of AlN sintered at 1800 °C.

The results of Differential Thermal Analysis and Thermogravimetry of AlN with the additive YF_3 (3.0 wt%) are shown in Figure 3. There is a small endothermic peak, and the weight change starts at about 700°C. After these analyses, the powder was checked by X-ray diffraction (Figure 4). From the X-ray diffraction (XRD) pattern it can be seen that YOF and $Y_3Al_5O_{12}$ were present below 950 ℃.

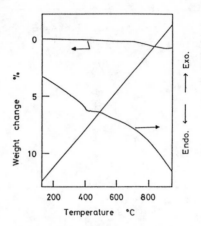

Figure 3. DTA-TG curves of AlN with YF_3.

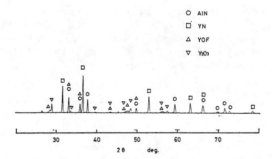

Figure 4. X-ray diffraction pattern after DTA-TG analysis.

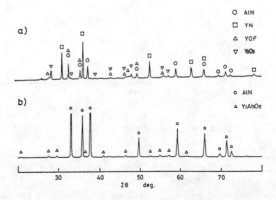

Figure 5. X-ray diffraction pattern of AlN sintered at 1800 ℃.

Figure 5 shows the XRD pattern of the sample containing YF_3 (3.0 wt%) sintered at 1800°C. Figure 5 (a) shows the XRD pattern for the AlN surface, and Figure 5 (b) for the AlN powder. YN is present on the surface of sintered AlN, and $Y_3Al_5O_{12}$ is present in the sintered AlN. This implies the following reaction.

$$Y_3Al_5O_{12} + 12C + 4N_2 \rightarrow 5AlN + 3YN + 12CO \qquad (1)$$

YF_3 reacts with the Al_2O_3 present on the surface of the raw AlN powder forming $Y_3Al_5O_{12}$. When Y_2O_3 is added, the formation of $Y_3Al_5O_{12}$ is due to the liquid phase sintering of AlN.

CONCLUSION

The effect of Y_2O_3 and YF_3 addition on the crystallographic and thermal properties of AlN is not so significant for samples sintered at 1800°C. But the quantity of liquid phase that is formed during sintering depends on the addition of Y_2O_3 and YF_3. However, for samples sintered at 1600°C, liquid phase sintering does not shorten the sintering time. At this temperature, dense samples can be obtained with a longer sintering time. The additive YF_3 reacts with Al_2O_3 at 950 °C as follows

$$Al_2O_3 + 3YF_3 \rightarrow 3YOF + 2AlF_3 \qquad (2)$$
$$3Al_2O_3 + 3YF_3 \rightarrow Y_3Al_5O_{12} + AlF_3 \qquad (3)$$

It can, therefore, be concluded that these additives trap oxygen and prevent it from dissolving the crystalline AlN.

REFERENCES

1. Komeya K., Inoue H., and Tuge A., Effect of various additives on sintering of aluminum nitride. Yogyo-Kyokai-Shi,1989,89[6],330-36.

2. Hazeyama I., Katakura H., and Ichinose N., Densification and thermal properties of low temperature firing AlN ceramics. presented at IEMT of Japan,1991,215-18.

3. Ichinose N. and Hazeyama I., Thermal properties of low temperature firing AlN ceramics, Report of Material Science and Technology (University of Waseda) 41,1991,Dec.,11-15.

4. Slack G.K., Nonmetallic crystals with high thermal conductivity. J.Phys.Chem.Solids,1973,34,321-335.

RECENT DEVELOPMENTS IN UNDERSTANDING GRAIN-BOUNDARY ELECTRIC AND DIELECTRIC PROPERTIES OF PTCR BARIUM TITANATE CERAMICS

DA YU WANG

GTE Laboratories Incorporated, Waltham, MA 02173, USA

and

KAZUMASA UMEYA

GTE Products Corporation, Standish, ME 04084, USA

ABSTRACT

Electrode PTCR effect was found on the surfaces of barium titanate ceramics. The electrodes were vacuum-deposited Mn with a Ag top-layer to enhance the conductance. In the current-voltage (IV) study, the electrodes showed a linear IV behavior at low voltages and an exponential type of diode behavior at high voltages. This IV characteristic could be explained by an equivalent circuit which had a restrictive resistance (ohmic contact) in parallel with a Schottky diode. In the temperature-IV study, no charge compensation effect of the spontaneous polarizations was found on either the ohmic contact or the Schottky diode. The PTCR characteristics of the Mn/Ag electrodes were similar to the PTCR effect of the bulk and were believed to be controlled by the bulk grain boundaries near the electrode areas.

INTRODUCTION

The positive-temperature-coefficient-resistor (PTCR) effect of barium titanate ceramics is a grain-boundary phenomenon [1]. Near the Curie point, the grain-boundary resistances jump several orders of magnitude, with the ferroelectric state having the lower resistance values. This effect can be enhanced by adding minute amounts of Cu, Fe, and Mn to the ceramics, and many important applications have been developed [2, 3].

The PTCR effect of the barium titanate ceramics was explained as an effect of a charge compensation mechanism of the spontaneous polarizations on the barrier heights at grain boundaries [4]. The ac complex impedance study indicated that the barium titanate grain boundaries could be treated as a double-Schottky barrier in which the dielectric constant was nonlinear and could be described by the Devonshire theory of barium titanate [3, 5-8].

Recently we studied the electrode polarization phenomena by using PTCR barium titanate ceramics as the substrate materials [9]. We found, in general, that the electrode low-polarization resistance also showed a PTCR effect, and the electrodes could be treated as single-phase or mixed-phase junctions, depending on the preparation procedures of the electrodes. The current-voltage (IV) and capacitance-voltage (CV) behaviors of single-phase electrodes could be described well by a Schottky barrier with a nonlinear dielectric constant described by the Devonshire theory of barium titanate. For mixed-phase electrodes, the IV and CV characteristics indicated that the electrode is either multi-Schottky barriers with different barrier heights or a Schottky barrier in parallel with a linear resistance $R_{//}$ [9, 10].

The origin of this parallel resistance $R_{//}$ could be an interesting subject of study. It may bring new understanding to the PTCR phenomenon of barium titanate ceramics and illustrate the charge compensation mechanism of the spontaneous polarizations on PTCR grain boundaries.

In this study, we researched such mixed-phase electrodes and studied their IV behaviors as well as the effect of the substrate materials on the electrode polarization mechanism.

THEORY

Based on the Schottky barrier model, the IV characteristic of a diode can be described as

$$I = I_o \exp(eV/nkT)[1 - \exp(-eV/kT)] \tag{1}$$

in which n is the diode ideality factor, and e, k, and T are the electron unit, the Boltzmann constant, and the temperature. Plotting the IV data in terms of Ln $I/[1 - \exp(-eV/kT)]$ vs V, a linear slope could be obtained, and the values of I_o and n can be determined from it [11].

Generally, I_o is a function of temperature. Depending on the transport mechanisms at the electrode junction, the pre-exponential constant I_o in Eq. (1) can have different expressions. When the transport is controlled by thermal diffusion, I_o is defined as

$$I_o = I_o{}' \exp[-(eV_n + eV_{bi})/kT]$$

with

$$I_o{}' = [e^3 D_n N_c N_D/(kTC)], \tag{2}$$

where D_n, N_c, N_D, and C are the electron diffusion constant, effective density of states in conduction band, donor impurity density, and electrode capacitance. The V_n is the potential difference between the edge of the conduction band and the Fermi level, and V_{bi} is the built-in potential.

If the transport is controlled by the thermionic emission, I_o is expressed as

$$I_o = I_o{}' \exp[-(eV_n + eV_{bi})/kT],$$

with

$$I_o{}' = A^* T^2, \tag{3}$$

in which A^* is the Richardson constant (11). Therefore, I_o is temperature-dependent, and if its value is known (from the IV experiments), the built-in potential can be estimated by using Eq. (2) or (3). Also based on Eq. (1), the low-voltage electrode resistance (with $V < kT/e$) can be derived as

$$R = (kT/eI_o). \tag{4}$$

Semiconductors can have mixed-phase contacts, and the above equations are still valid under such situations except that the final IV equations will depend on the structure of the equivalent circuit and change accordingly [12].

EXPERIMENTAL

Sample Preparation — The barium titanate samples studied were prepared by the standard solid state reaction method (5). The basic starting materials were $BaCO_3$ and TiO_2 powders. Y_2O_3 was added (0.25 mole %) to impart the semiconducting properties. Extra TiO_2 and SiO_2 (0.5%) were added to improve sintering. $MnNO_3$ (150 ppm) was added to enhance the PTCR effect. To study the effect of the substrate materials on the electrode polarization phenomena, part of the samples were prepared without $MnNO_3$ as the additive, and the content of Y_2O_3 was adjusted so that the samples have a room-temperature bulk (lattice and grain-boundary) resistance value similar to that of the samples containing $MnNO_3$.

Typical samples had a thickness of 1.0 mm and a diameter of 13.8 mm. Powders with correct weight proportions were first milled overnight, calcined at 1000°C for 1 hr, milled again, and then cold-pressed into pellets. All the sintering was done at 1360°C for a half hour in air. The sample densities were between 5.30 and 5.40 g-cm^{-3}. The average grain size of the samples was 12 mm.

The samples thus prepared had typical resistance between 0.7 and 0.9 W at room temperature with In-Ga, a well-known ohmic contact material for barium titanate, as the electrodes [13]. The dielectric properties of the samples were determined by studying samples purposely prepared without the addition of yttrium. Typical temperature-dielectric-constant data published in Reference 8 gave a Curie temperature at 127.3°C, a Curie point at 135°C, and a Curie constant of 1.1×10^5°C. The effective donor level of our ceramic samples was estimated to be $N_D = 5 \times 10^{18}$ cm^{-3}, based on the lattice resistivity values (the grain-boundary resistance values were not included).

All the electrode samples studied in this work had one side coated with vacuum-deposited Mn, which was covered by another layer of Ag for better conductance, and the other side coated with In-Ga. Both electrodes had an edge-to-edge resistance value between 0.2 and 0.3 W.

When the Mn/Ag electrode samples were subjected to an air-heat treatment, the Mn/Ag electrodes were put against a smooth alumina ceramic surface. After the heat-treatment, the Mn/Ag electrodes were checked again for the edge-to-edge resistance, which usually did not change much from its starting value.

Electrical Measurement — In the IV measurement, the In-Ga electrode and the Mn/Ag electrode were pressed lightly against gold foils (0.1 mm thick and 13.8 mm in diameter) which were connected to the power supply (Keithley 225 Current Source) and digital multimeters. The IV measurements were done in an air furnace with the temperature controlled to within 0.1°C. The bulk (lattice and grain-boundary) contributions to the polarizations were subtracted from the IV data by using the resistance data determined by the ac complex impedance method [5]. Two types of equipment were used for this purpose: an HP-4192A LF Impedance Analyzer and a Solartron-1250 Frequency Response Analyzer (together with a Solartron-1186 Electrochemical Interface). The covered frequency range was 10 Hz to 13 MHz for the HP-4092 and 0.0001 Hz to 64 kHz for Solartron-1250. The amplitude of the a–c signals used was 10 mV.

RESULTS

All the Mn/Ag electrodes we studied formed ohmic contact with the substrate materials after the vacuum deposition procedures (this was confirmed by the ac complex impedance method; no electrode arc was measured). When the Mn/Ag electrodes were exposed to air-heat treatments, the electrode PTCR characteristics developed. Typical results of such electrode samples are given in Figure 1.

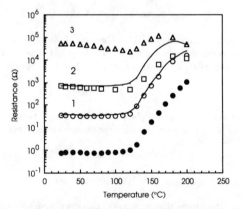

Figure 1. The plots of the temperature-electrode-resistance data of three Mn/Ag electrode sample, the dark circles represent the bult resistance values. Sample 1 (open circles) was heated to 450°C in air for 30 min, sample 2 (open squares) to 550°C for 30 min, and sample 3 (open triangles) to 650°C for 30 min.

In Figure 1, three sets of electrode resistance data are plotted as Log R vs T. The data belonged to three samples: sample 1 (open circles) was heated to 450°C in air for 30 min, sample 2 (open squares) to 550°C for 30 min, and sample 3 (open triangles) to 650°C for 30 min. These electrode resistance data were measured at low voltage (<20 mV). Also shown in Figure 1 are the bulk (lattice and grain-boundary) resistance data of a typical ceramic sample (dark circles) which had In-Ga for both electrodes. As clearly shown in Figure 1, the electrode resistance data had PTCR effect, and the resistance level depended on the sample heat-treatment temperatures.

In general, we found that the IV characteristics of Mn/Ag electrodes had linear IV behaviors at low voltages and exponential-type IV behaviors at high voltages. Typical electrode IV characteristics are shown in Figures 2 and 3, plotted as Log I vs V (solid circles in Figures 2 and 3). The data belonged to sample 1 in Figure 1, and the temperature was 24°C and 160°C, respectively. The IV results could be described by an equivalent circuit which was composed of a Schottky barrier (described by Eq. [1]) in parallel with a linear resistance $R_{//}$. The solid lines in Figures 2 and 3 represented the individual fitting result of $R_{//}$, Eq. [1], and the combination of the two (the constants used in the fitting are listed in Table 1).

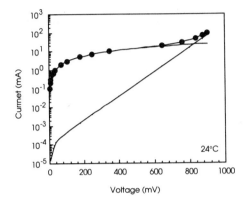

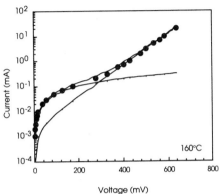

Figure 2. The IV plots of sample 1 at 24°C.

Figure 3. The IV plots of sample 1 at 160°C.

TABLE 1
Constants obtained from fitting the IV data of samples 1, 2, and 3

T(°C)	R//	I_o (mA) Sample 1	n	R//	I_o (mA) Sample 2	n	R//	I_o (mA) Sample 3	n
24	35.2	1.1×10^{-4}	2.60	625	1.4×10^{-4}	2.47	464000	8.1×10^{-4}	2.66
160	2066	2.1×10^{-3}	1.85	10341	3.7×10^{-3}	2.35		5.3×10^{-4}	1.78
170	4650	1.5×10^{-4}	1.80	26210	2.1×10^{-3}	1.81		5.1×10^{-4}	1.67

When the electrode resistance became large, the diode IV characteristic became the dominant part of the IV data. A typical example can be seen in Figure 4, in which the IV data (solid circles) of sample 3 at 160°C are plotted as Ln I/[1 − exp (−eV/kT)] vs V. A linear slope (see the solid line in Figure 4) can be obtained from the plots, which indicates that Eq. (1) alone is sufficient to describe the IV behavior of the data.

For easy comparison, the fitting results (the values of $R_{//}$, I_o, and n) of samples 1, 2, and 3 are shown at three different temperatures (24°C, 160°C, and 170°C) in Table 1. Table 1 clearly shows that at least for samples 1 and 2, the R// values did not disappear at high temperatures (> the Curie point); the three samples had similar n values which decreased as the temperature increased; and the three samples had similar I_o values (at least within one order of magnitude) but not with the $R_{//}$. Based on Eqs. (2) and (3) and the I_o data shown in Table 1, the barrier height of the Schottky barrier at room temperature is not much smaller than that at a temperature higher than the Curie point.

Two depressed arcs were observed in the ac complex impedance measurements of these Mn/Ag electrode samples, with a high-frequency arc corresponding to the lattice resistance and capacitance (PTCR grain boundaries) and a low-frequency arc corresponding to the electrode resistance and capacitance. The electrode resistance values determined by both small-signal ac and small-voltage dc methods agreed well with each other.

Since the parallel circuit model fits the electrode IV data well (see Figures 2–4), we use the same idea in fitting the temperature-electrode-resistance data shown in Figure 1. We assume the parallel $R_{//}$ is related to the bulk (lattice and grain-boundary) resistance (dark circles in Figure 1) by a constant G, and the electrode-resistance R could be expressed as

$$R = 1 / [(G/RB) + (eI_o/kT)]. \tag{5}$$

Since the IV data of sample 3 (see Figures 1 and 4 and Table 1) are dominated by the diode IV behavior, we use sample 3's data, shown in Figure 1 (the open triangles), and substitute into Eq. (5) to fit the data of samples 1 and 2, shown in Figure 1. The solid curves in Figure 1 represented the fitting results with a G value of 50 and 926, respectively. As shown in Figure 1, the fitting results are good, especially for sample 2.

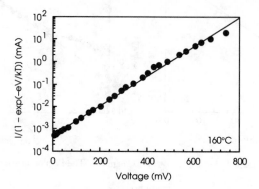

Figure 4. The IV plots of sample 3 at 160°C.

To study the effect of the substrate materials on the electrode polarization behaviors, as shown in Eq. (5), we prepared electrode samples by using ceramic samples without Mn as the additive. Ceramic samples without Mn as the additive would have a suppressed grain-boundary PTCR effect. In Figure 5, typical bulk (lattice and grain-boundary) resistance data of such prepared ceramic samples (represented by the solid circles), obtained by using In-Ga electrodes, are given. Compared with the solid circle data shown in Figure 1, the PTCR effect shown in Figure 5 is less by one order of magnitude.

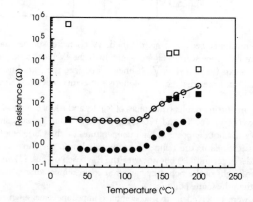

Figure 5. The plots of the temperature-electrode-resistance data of one Mn/Ag electrode sample (substrate contained no MN); the dark circles represent the bulk resistance values.

Like the samples prepared on ceramics with Mn added, the Mn/Ag electrodes formed ohmic contact with the new substrate materials. Only after heat treatment in air did the electrode samples show PTCR effect. A typical example of the electrode PTCR resistance data is given in Figure 5 for a Mn/Ag sample (open circles) which had been heated to 450°C in air for 30 min.

The IV characteristics of this Mn/Ag sample were similar to those shown in Figures 2 and 3, with the data fitted well by a parallel circuit of a linear R// and a Schottky diode. A typical example could be seen in Figure 6, in which the IV data (dark circles) were plotted as Log I vs V. The solid lines in Figure 6 represent the fitted results. In Table 2, we listed the fitting results of the same sample at four different temperatures (24°C, 160°C, 170°C, and 200°C). Also listed are the resistance data calculated based on Eq. [4] (both data of $R_{//}$ and kT/eI_o in Table 2 are plotted in Figure 5 also, with dark squares for $R_{//}$ data and open squares for data of kT/eI_o). Compared with Table 1, Table 2 gives smaller high-temperature values for $R_{//}$, and the rest of the data are similar.

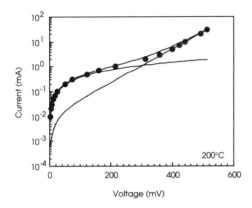

Figure 6. The 200°C IV plots of the sample introduced in Figure 5.

TABLE 2
Constants obtained from the fitting results shown in Figures 4 and 5

Temperature (°C)	R// (Ω)	I_o (mA)	n	kT/eI_o (Ω)
24	16	5.1×10^{-5}	2.05	5.1×10^5
160	160	1.7×10^{-3}	1.64	2.3×10^4
170	180	1.6×10^{-3}	1.56	2.4×10^4
200	268	1.0×10^{-2}	1.51	3.9×10^3

We use Eq. (5) to fit the temperature-electrode-resistance data shown in Figure 5. The bulk (lattice and grain-boundary) resistance data in Figure 5 (see the dark circles in Figure 5) are used for R_B in Eq. (5), and the contributions of the diode term in Eq. [5] are neglected. The solid curve in Figure 5 represents the fitted results. Other than the high-temperature part of the data, the fittings are good.

DISCUSSION

We suspect the parallel resistance $R_{//}$ observed in the electrode IV study was the restrictive resistance of an ohmic contact due to the slow mobility of the electron carriers in the substrate materials. This statement was supported by the observation that the electrode PTCR effect follows the change of the bulk (lattice and grain-boundary) PTCR effect (see Figures 1 and 5).

The dependence of restrictive resistance on the bulk (lattice and grain-boundary) resistivity is usually linear, and the geometrical factor G in Eq. (5) could be a complex function of the dimensions of the sample and electrode structure. Unless the electric current distributions are changed due to the change of the electrode structures at different temperatures, the G factor should be temperature independent (14). Therefore

Eq. (5) should produce reasonable fitting results for the data shown in Figures 1 and 5. In Figure 1, the high temperature data of sample 2 were one order of magnitude smaller than the predicted values (see the solid curve and open squares in Figure 1). This could be explained by the fact that the actual I_o values of sample 2 were actually one order of magnitude higher than that of sample 3, of which the I_o values were used in the fittings. The deviations observed in Figure 5 (between the fitting curve and the solid squares) were real and could not be explained by the variation of the bulk (lattice and grain-boundary) resistance between the two samples. Since only one arc was observed for the electrode polarization mechanism in the small-signal ac complex impedance study, the restrictive resistance was viewed as a local resistance developed between the electrode area of Schottky diode and that of the ohmic contact area. In this view, the $R_{//}$ represents the resistance value of the substrate material which is near the surface where the electrode is located. This part of ceramic might have a slightly different resistance value from that of the bulk (lattice and grain-boundary). Also the restrictive current running between the two areas of different electrodes (ohmic contact and Schottky barrier) should pass through several grain boundaries in order to inherit the PTCR phenomenon.

The most surprising result from this study is that there is no solid evidence to show that the charge compensation effect of the spontaneous polarizations worked on the electrode. This statement was supported by the fact that the existence of $R_{//}$ was observed at a temperature higher than the Curie point (see Figure 6) and that the low-temperature I_o values listed in Tables 1 and 2 were higher than those measured at high temperatures. If the charge compensation mechanism works here, one would expect, once the temperature passes the Curie point, a room-temperature ohmic contact would become a Schottky diode (therefore a disappearance of the $R_{//}$), and a room-temperature Schottky diode would increase its barrier height and decrease its I_o value dramatically [4].

The absence of $R_{//}$ for sample 3 in Table 1 should not be taken as evidence of the disappearance of the ohmic contact. The correct explanation for not seeing it in the IV measurement is that the restrictive resistance value was too high and the electrode IV characteristics are dominated by the more conductive Schottky barrier.

The diode ideality factor n observed here is larger than that reported in the literature for an electrode on semiconductors such as Si and III-V compound semiconductors. Several mechanisms have been cited in the literature for explaining why the diode ideality factor is larger than one: (a) the effect of image-force lowering on the barrier height, (b) field or thermionic field emission effect of the transport mechanism on the forward characteristic, (c) electron-hole recombination in the depletion region (in this case, n=2), and (d) the effect of an interfacial layer [15]. We believe the last model is the most plausible one in explaining our results.

ACKNOWLEDGMENTS

Special thanks to Dr. B. Ditchek of GTE Laboratories Incorporated for encouraging us to publish this work.

REFERENCES

1. Haayman, P.W., Dam, R.W., and H.A. Klasens, "Method of Preparation of Semiconducting Materials," German Patent 929,350, June 23, 1955; British Patent No. 714,965.

2. Jaffe, B., Cook, W.R., and Jaffe, H., *Piezoelectric Ceramics*, Academic Press, New York, 1971, pp. 53-115.

3. Kulwicki, B.M., PTC Materials Technology, 1955-1980. In *Advances in Ceramics*, Vol. 1, Grain Boundary Phenomena in Electronic Ceramics, ed. L.M. Levinson and D.C. Hill, American Ceramic Society, Columbus, OH, 1981, pp. 138-154.

4. Heywang, W., Barium Titanate as a Semiconductor with Blocking Layers. *Solid-State Electron.*, 1966, **3** [1] pp. 51-58. Jonker, G.H., Some Aspects of Semiconducting Barium Titanate. *Solid-State Electron.*, 1964, **7**, pp. 895-903.

5. Wang, D.Y. and Umeya, K., Electrical Properties of PTCR Barium Titanate. *J. Am. Ceram. Soc.* 1990, **73** [3], pp. 669-7.

6. Wang, D.Y. and Umeya, K., Depletion-Layer Dielectric Properties of PTCR Barium Titanate. *J. Am. Ceram. Soc.* 1990, **73** [6], pp. 1574-81.

7. Wang, D.Y. and Umeya, K., Spontaneous Polarization Screening Effect and Trap State Density at Grain Boundary of Semiconducting Barium Titanate Ceramic. *J. Am. Ceram. Soc.* 1991, **74** [2], pp. 280-6.

8. Devonshire, A.F. and Wills, H.H., Theory of Ferroelectrics. *Adv. Phys.* 1954, 3 [10], pp. 85-130.

9. Wang, D.Y.,Surface and Bulk PTCR Effects of Ferroelectric Barium Titanate Ceramics. Vol. 24 of Ceramic Transactions, Am. Ceram. Soc., 1991, pp. 349-57.

10. Wang, D.Y., Electrode Contact Properties for Semiconducting Ferroelectric Ceramics. 59-E-90, 92 Annual ACS Meeting, April 22-26, Dallas, TX. (Paper has been submitted to *J. Am. Ceram. Soc.*)

11. Rhoderick, E.H. and Williams, R.H., *Metal Semiconductor Contacts*, Clarendon Press, Oxford, 1988, p. 99.

12. Ohdomari, I. and Tu, K.N., Parallel Silicide Contacts. *J. Appl. Phys.* 1980, **51** (7), pp. 3735-9.

13. Sauer, H.A. and Flaschen, S.S., Choice of Electrodes in Study and Use of Ceramic Semiconducting Oxides. *Ceramic Bulletin* 1960, **39** (6) 304-6.

14. Smits, F.M., Measurement of Sheet Resistivities with the Four-Point Probes. *J. Bell System Tech* 1958, **37**, pp. 711-18.

15. Rhoderick, E.H. and Williams, R.H., *Metal Semiconductor Contacts*, Clarendon Press, Oxford, 1988, p. 133.

GRAIN BOUNDARY PROPERTIES OF OXYGEN HOT ISOSTATICALLY PRESSED BARIUM STRONTIUM TITANATE WITH POSITIVE TEMPERATURE COEFFICIENT OF RESISTIVITY

BEN HUYBRECHTS, KOZO ISHIZAKI AND MASASUKE TAKATA

Nagaoka University of Technology

Niigata 940–21, Japan

ABSTRACT

The influence of annealing at high oxygen partial pressures (O_2–HIP) on the Positive Temperature Coefficient of Resistivity (PTCR)–behavior of Mn–doped $Ba_{0.8}Sr_{0.2}TiO_3$ was examined. Although no significant microstructural or density changes could be observed as a consequence of the O_2–HIP annealing, the electrical resistivity characteristics changed markedly. The maximum resistivity was increased about 160 times and the temperature of maximum resistivity decreased from 209°C to 149°C. The grain resistivity is not changed, which shows that O_2–HIPping only modifies the grain boundary structure. From the Heywang model and the gradient in the Arrhenius plot i.e., resistivity as a function of the reciprocal of the temperature, an increase of 44% in the acceptor state density was estimated.

INTRODUCTION

It has been known since 1955 that the resistivity of polycrystalline n–type $BaTiO_3$ increases very rapidly in a narrow region above the tetragonal–cubic transformation temperature, 130°C. The most accepted model to explain this Positive Temperature Coefficient (PTC) is the Heywang model first published in 1961 [1]. Heywang assumed a two–dimensional layer of acceptor states at the grain boundaries. These acceptors attract electrons from the bulk, resulting in the build–up of a potential barrier. Several attempts have been made to increase both the positive temperature coefficient of the resistivity and the resistivity–jump [2–6]. The addition of Mn has been reported to be very effective [2–5].

Small additions of Mn can improve the PTC–behavior by increasing the number of acceptors [2–4], and by forming deeper acceptor traps. The following mechanism is proposed to explain this improvement : Due to the reduction of Mn to Mn^{2+} at high temperatures, Mn is present as Mn^{2+} during the sintering period. During cooling the Mn^{2+}, which is a donor, is oxidized to higher oxidation states [2,7] and forms deep acceptor levels [8,9]. From this mechanism it is clear that the oxidation of the Mn ions is crucial for obtaining steep and large jumps. It was therefore suggested that annealing PTC–material under high oxygen pressures should increase the number of Mn ions in the highest oxidation states thereby improving the PTC–behavior.

Although the effect of high oxygen partial pressures on acceptor doped $BaTiO_3$, has been studied by Hagemann et al. [10], as far as we can ascertain it has not yet been applied to PTC–material. To investigate the effect of high oxygen partial pressures on PTC–materials, $Ba_{0.8}Sr_{0.2}TiO_3$–samples were annealed in an O_2–HIP.

EXPERIMENTAL

The $Ba_{0.8}Sr_{0.2}TiO_3$ was produced by a conventional solid–state reaction of a mixture of $BaCO_3$, $SrCO_3$ and TiO_2 in a mol–ratio of 80:20:101. The reactants were ball–milled in ethanol for 16 hours, using ZrO_2 balls in a nylon vessel. The powder was then dried, followed by presintering at 1100°C for 3 hours in air. After crushing and sieving the reacted powder, dopants (0.15 mol% Sb_2O_3 and 0.04 mol% $MnCO_3$) and additives (1.5 mol% SiO_2 and 1 mol% Al_2O_3) were ball–milled (nylon vessel, ZrO_2–balls) in ethanol for 16 hours. After spray–drying, the powder mixture was uniaxially pressed (100 MPa) into pellets of a diameter of 20 mm. The pellets were sintered for 1 hour at 1320°C. Post–treatment in O_2–HIP took place under P_{tot}=100 MPa (90% Ar, 10% O_2) and P_{O2}=10 MPa at T=1200°C, for 1 hour.

The resistivity was measured after In–Ga electroding. The log ρ–T curves were obtained using a computer controlled 2–probe technique. Complex–impedance measurements in the 5Hz–13Mhz range, were used to determine the grain resistivity. Details are described elsewhere [11]. The density was calculated according to Archimedes' method after weighing in air and in water. The grain size was obtained by an intercept technique.

RESULTS

The O_2–HIP treatment had a negligible effect on the density and microstructure, as can be seen in figure 1. The electrical properties, on the other hand, changed markedly. Figure 2 shows the large difference in resistivity behavior before and after O_2–HIP annealing. The resistivity at 30°C, ρ_{30}, rises by a factor of 10, but the maximum resistivity, ρ_{max}, rises by a factor of 160. Figure 2b shows that S_{arrh}, the opposite sign of the slope in the Arrhenius plot is increased by a factor of 2. The data are summarized in table 1.

TABLE 1
Properties as sintered and after O_2-HIP

Property	As sintered	After O_2-HIP
Grain size	5 μm	6 μm
Density	5.65 g/cm^3	5.76 g/cm^3
ρ_{max}	1.26 MΩcm	199 MΩcm
T_{max}	209 °C	149 °C
ρ_{min}	45 Ωcm	451 Ωcm
S_{arrh}	6 900 K	14 400 K
ρ_0	4 Ωcm	4 Ωcm

DISCUSSION

As previously stated, Heywang assumed a two–dimensional layer of acceptor–states at the grain boundaries. These acceptor–states attract electrons from the bulk, resulting in a build-up of a potential barrier, ϕ_0 :

$$\phi_0 = \frac{eN_s^2}{8e_0 e_r N_d} \tag{1}$$

where N_s is the acceptor–state density, N_d is the charge carrier density, e is the electron

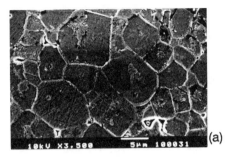 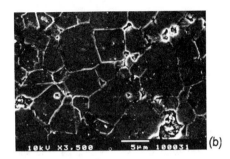

Figure 1. The microstructure after polishing and etching. (a) as sintered (b) after O_2–HIP. No increase in grain size could be observed. The scalemarker indicates 5 μm.

charge and ε_r is the relative permittivity at the grain boundary.

The temperature dependence of ε_r obeys the Curie–Weiss law :

$$\varepsilon_r = \frac{C}{T-\theta} \qquad (2)$$

where C represents the Curie constant, θ the extrapolated Curie–Weiss temperature, and T the absolute temperature.

The resistivity of the PTC–samples is related to the potential barrier by the following equation :

$$\rho = \alpha \rho_o \exp(\frac{e\phi_0}{kT}) \qquad (3)$$

where α is a geometrical factor, and ρ_0 is the bulk resistivity. It is clear that the resistivity increases steeply above the Curie–point, T_c, because of the sudden drop of ε_r. In addition, with increasing temperature the acceptor–states are raised together with the conduction band, until they reach the Fermi–level. At the Fermi–level, trapped electrons start to jump to the conduction band. This results in depressing the increase in resistivity, and ultimately in enhancing the conductivity.

From the Heywang model it follows that an increase in acceptor–states results in an increase in ρ_{max}, and a decrease in T_{max}. Returning to the present data, changes can be seen in ρ_{max} and T_{max}. This behavior can be explained by an increase in N_s. Because in the vicinity of the maximum resistivity the number of filled acceptor–states is different from the total number of acceptor–states, the calculation of N_s from ρ_{max} data has to be done iteratively [1,12]. However, we can also calculate the increase in N_s using the parameter S_{arrh}. Compared to the maximum resistivity, S_{arrh} is independent of the grain size, the energy level of the acceptor–states, and the sample dimensions.

From equations 1, 2, and 3 , it can be shown that S_{arrh} is proportional to N_s^2/N_d[13,14]. The charge carrier density, N_d, is obtained from complex–impedance measurements. Figure 3 shows that the grain resistivity, which is the high frequency intercept with the X– abscissa, is the same before and after O_2–HIP treatment. This is evidence that the treatment only influenced the grain boundary and not the grain itself. A change in S_{arrh} can, therefore, only be explained by a change in N_s. The increase in N_s can be estimated from the ratio of S_{arrh}. Using the data in table 1, we find that O_2–HIPping resulted in a 44% increase in N_s.

The increase in the low temperature resistivity, ρ_{30}, can be explained by the model of Jonker, which is based on the ferro–electric behavior of $BaTiO_3$ below the Curie temperature [12]. Jonker postulated that the acceptor–states at lower temperatures are compensated by a difference in the direction of polarization between two adjacent grains. There is however a critical number of acceptor–states that can be compensated [13,15]. If the number of acceptors is larger than the maximum number of acceptors that can be compensated, the low temperature resistivity increases with increasing N_s. This explains the increase in ρ_{30} after O_2–HIP treatment.

Thus far we have shown that O_2–HIP treatment results in an increase in acceptor–

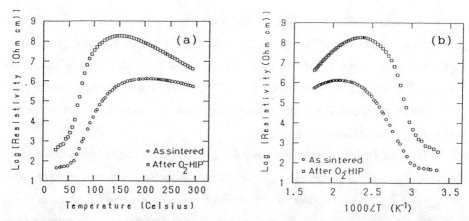

Figure 2. (a) The log ρ–T curve for a sample in the as sintered state and after the O_2–HIP treatment. Note the increase in ρ_{min} and ρ_{max}, and the decrease in T_{max}. (b) The Arrhenius plot, log ρ – 1/T indicates that the slope became sharper after O_2–HIP treatment.

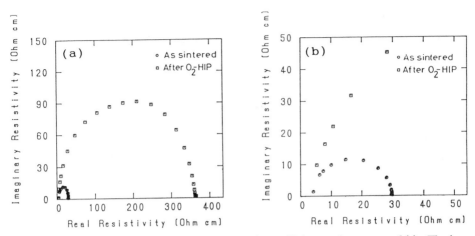

Figure 3. Complex–impedance measurements. Figure (b) is an enlargement of (a). The low frequency intercept with the x–abscissa is the dc–resistivity. The high frequency intercept is the grain resistivity. It is clear that although the dc–resistivity increases by a factor 10, the grain resistivity does not change after O_2–HIP.

state density at the grain boundaries. Although the exact mechanism for the increase is not completely clear, we believe that it can be explained in two ways. First, it is known that the additional charge resulting from the incorporation of Mn at the Ti–sites in the lattice is compensated by double–ionized oxygen vacancies [10,16]. It has also been postulated earlier that O_2–annealing at temperatures below the sintering temperature decreases the number of oxygen vacancies. This leads to a higher effective Mn–concentration and hence to a higher N_s [9]. A similar mechanism can take place during O_2–HIP treatment. A second mechanism that can not be disregarded is the segregation of Mn at the grain boundaries. Possible acceptor segregation in $BaTiO_3$ has been shown by Desu et al. [17] and Heydrich et al.[18]. A combination of both mechanisms is another possibility.

CONCLUSIONS

In conclusion, O_2–HIPping had a negligible effect on the density and microstructure. Furthermore it did not influence the grain resistivity, showing that only the grain boundaries are affected by this treatment. The increase in ρ_{max}, ρ_{30}, S_{arrh}, and the decrease in T_{max} after O_2–HIPping can be explained by an increase in N_s. From the ratio of S_{arrh}, we estimated that this increase is 44 %. It is suggested that the 44% increase in N_s is caused by an increase in effective Mn–concentration due to a decrease in double ionized oxygen vacancies. However, segregation of Mn to the grain boundaries during O_2–HIP could also cause an increase in the density of acceptor–states at the grain boundaries.

ACKNOWLEDGEMENTS

We thank INAX for providing the as sintered samples and M. Kinemuchi for the help in measuring and analyzing. B. Huybrechts also thanks the European Community for the *"Scientific Training Programme in Japan"*–fellowship, which made this research possible.

REFERENCES

1. Heywang, W., Bariumtitanat als Sperrschichthalbleiter, Sol. State Elec., 1961, 3, pp.51

2. Ueoka H., The Doping Effects of Transition Elements on the PTC Anomaly of Semiconductive Ferroelectric Ceramics, Ferroelectrics, 1974, 7, pp.351

3. Ihrig H., PTC Effect in $BaTiO_3$ as a Function of Doping with 3d–Elements, J. Am. Ceram. Soc., 1981, 64, 16, pp.617

4. Illingsworth J., Al–Allak H.M., Brinkman A.W., Woods J., The influence of Mn on the Grain–Boundary Potential Barrier Characteristics of Donor Doped $BaTiO_3$ Ceramics, J. Appl. Phys., 1990, 97, 4, pp.2088

5. Matsuoka T., Matsuo Y., Sasaki H., Hayakawa, PTCR Behavior of $BaTiO_3$ with Nb_2O_5 and MnO_2 Additives, J. Am. Ceram. Soc., 1972, 55, 2, pp.108

6. Jonker G.H., Halogen Treatment of Barium Titanate Semiconductors, Mat. Res. Bull., 1967, 2, pp.401

7. Ting C.T., Peng C.J., Lu H.Y. and Wu S.T., Lanthanum Magnesium and Lanthanum Manganese Donor–Acceptor–Codoped Semiconducting Barium Titanate, J. Am. Ceram. Soc., 1990, 73 , 2, pp.329

8. Nakahara M., Murakami T., Electronic States of Mn ions in $Ba_{0.97}Sr_{0.03}TiO_3$ single crystals, J. Appl. Phys., 1974, 45, 9, pp.3795

9. Al–Allak, H.M., Brinkmann, A.W., Russell, G.J. and Woods, J., The Effect of Mn on the Positive Temperature Coefficient of Resistivity characteristics of donor doped $BaTiO_3$–ceramics, J. Appl. Phys., 1988, 63, pp.4530

10. Hagemann, H.J. and Ihrig, H., Valence Change and Phase Stability of 3d–doped $BaTiO_3$ annealed in oxygen and hydrogen, Phys. Rev. B, 1979, 20, pp.3871

11. Maiti, H.S. and Basu, R.N., Complex Plane Impedance analysis for Semiconducting Barium Titanate, Mat. Res. Bull, 1986, 21, pp.1107

12. Jonker, G.M., Some Aspects of Semiconducting Barium Titanate, Sol. State Elec., 1964, 7, pp.895

13. Huybrechts, B., Ishizaki, K. and Takada, M.,Experimental Evaluation of Acceptor-states Compensation in PTC–type Barium Titanate, accepted by the J. Am. Ceram. Soc.

14. Huybrechts, B., Ishizaki, K. and Takada, M., Influence of High Oxygen Partial Pressure on the Positive Temperature Coefficient Resistivity of $BaTiO_3$, to be published in Gas Pressure Effects on Materials, Processing and Design, ed. K. Ishizaki, Material Research Society Proceedings Series, Volume 251

15. Jonker, G.H., in Advances in Ceramics vol 1, ed. L.M. Levinson, Am. Ceram. Soc., Columbus, Ohio, 1981, pp.155

16. Hagemann, H.J. and Hennings, D., Reversible Weight Change of Acceptor Doped $BaTiO_3$, J. Am. Ceram. Soc., 1981, 64, pp.590

17. Desu, S.B. and Payne, D.A., Interfacial Segregation in Perovskites : Part II Experimental Evidence, J. Am. Ceram. Soc., 1990, 73, pp.3398

18. Heydrich, H. and Knauer, U., Grain Boundary effects in Ferro-electric Barium Titanate, Ferroelectrics, 1981, 31, pp.151

GAS DETECTION UTILIZING
THE INTERFACE OF A CuO/ZnO THIN FILM

YOSHIJIRO USHIO, MASARU MIYAYAMA*
and HIROAKI YANAGIDA**
NIKON Corp, Shinagawa-ku, Tokyo 140
*Research Center for Advanced Science and Technology,
University of Tokyo, Meguro-ku, Tokyo 153
**Faculty of Engineering,
University of Tokyo, Bunkyo-ku, Tokyo 113

ABSTRACT

The gas sensing function of the interface of CuO (p-type
semiconductor) and ZnO (n-type semiconductor) thin films was
investigated. Films were prepared by sputtering and the
junction showed rectified I-V characteristics originating from
the interface of the semiconductors. An array pattern was made
in the films (channels of ~10 μm were formed) by
photolithography so that the interfaces of the semiconductors
could be exposed to the atmosphere. These exposed junctions
showed current increases at a p-n forward bias when flammable
gases (CO, H$_2$) were introduced. In the temperature range from
100℃ to 250℃, the current increase was observed at 0.5 to
1.0 V bias, and at room temperature it occurred at a higher
bias than 2.0 V. Little change was observed at a p-n reversed
bias. The sensitivity (ratio of the current with and without
flammable gas) and gas selectivity were different depending on
the applied bias and the heat treatment temperature.

INTRODUCTION

It is well known that the interactions between different
materials can assume important roles in the functioning of

chemical sensors. However, the mechanisms are not yet completely understood, and there have been many investigations concerned with combining different materials to improve the sensing properties of chemical sensors.

There are also sensors that utilize interactions at the interfaces more directly by forming junctions or contacts between different materials. For these sensors, it is essential that the detecting molecules adsorb or permeate to the vicinity of the interfaces. In the diode type or MOS type sensors, porous electrodes and thin film sensing materials are used for accomplishing this desired result. [1] Hetero-contact type sensors are made by contacting sintered semiconductors that have rough surfaces initially. They have been found to possess unique sensing properties such as good gas selectivity and voltage-dependent sensitivities. [2~6]

In the present study, taking into consideration the poor reliability and reproducibility of the mechanically pressed hetero-contact type sensors, thin film CuO/ZnO interfaces joined and exposed to the atmosphere were fabricated, and their sensing properties for flammable gas (CO, H_2) were investigated.

MATERIALS AND METHODS

The semiconductive thin films used in this study were a CuO (p-type semiconductor ~0.5 μm) and a ZnO (n-type semiconductor ~0.7 μm). These films were prepared by sputtering targets of the sintered materials. Array patterns (10 μm pitch and 10 μm wide) were made by a photolithography process described elsewhere [7,8].

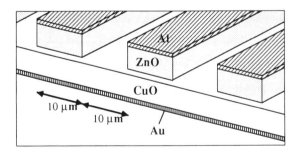

Figure 1. The structure of the thin film junctions

An Al film (~0.1 μm) and a Au film (~0.1 μm) were used as the

electrodes on the ZnO and CuO film respectively, to avoid
Schottky barrier formation at the electrode/semiconductor
interface. The structure of the thin film junctions is shown in
Fig.1. The junctions were placed in a temperature controlled
cell, about 600 ml in volume, and the current-voltage(I-V)
characteristics were measured when gaseous CO and H2 (~4000 ppm)
were introduced with dry air as the carrier gas. The total flow
rate was fixed at 100 ml/min. Measurements were made from room
temperature(~25℃) to 250℃.

RESULTS

The junctions showed rectified I-V characteristics attributed
to the CuO/ZnO interfaces. In the high humidity condition at
room temperature, only the current at the forward bias of the
p-n junctions (polarity of CuO+, ZnO-) increased. As a result,
the apparent rectification became significant, as shown in
Fig.2. [7,8]

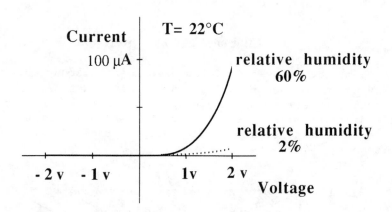

Figure 2. I-V Characteristics of the thin film junctions

When flammable gases were introduced, the current at the
forward bias increased. Figure 3 shows the current increase
when CO and H2 gases are introduced at a forward bias of 0.5 V.
The current increase was defined as $(I - I_0)/I_0$, where I is the
current five minutes after introduction of the flammable gases,
and I_0 is the current in dry air. The thin film junctions
without patterns, in which the vicinities of interfaces were

not exposed to the atmosphere, showed only small changes in the
current caused by the flammable gases. A marked difference of
sensitivity for CO and H2 gas was observed at about 150℃.

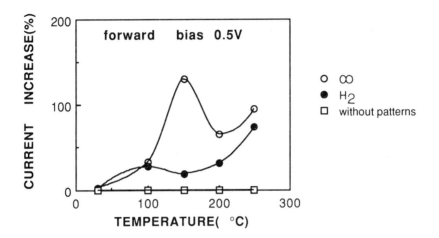

Figure 3.Current increase for CO(○) and H2(●)-flow at various
temperatures. □; response of unpatterned junction.

Even at room temperature (~25℃), increases in the current were
observed.

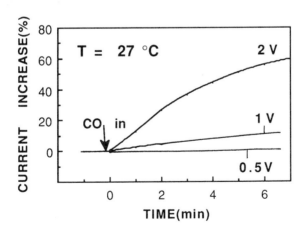

Figure 4.Current increases caused by CO gas at room temperature
at various forward biases.
As was seen in the humidity sensing measurement, the current

changes were small at a forward bias of 0.5 or 1 V and noticeable at 2 V.(Fig.4) In any case, the current changes at reversed bias were negligibly small.

Figure 5 shows the current increases at various temperatures at forward biases of 0.5 and 1.0 V. At temperatures above 150℃, a high applied bias did not always lead to a high sensitivity. When the bias was applied at higher temperatures, the current values sometimes fluctuated rapidly, making it difficult to measure the sensing properties.

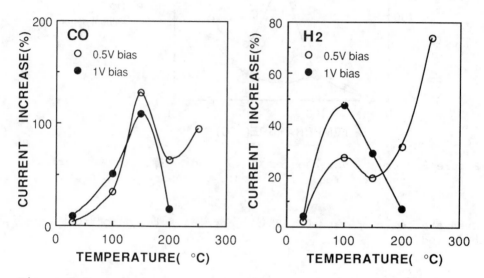

Figure 5.Current increases at a forward bias of 0.5V(○) and 1V(●) at various temperatures for CO and H_2-flow.

The current values at a forward bias of 0.5 or 1 V in dry air were stable. However, when fairly high forward biases were applied, for example 2 V at 200℃, the current sometimes continued to increase for several hours. These continuous increases were not observed at reversed bias. After increasing, the value of the current became stable while a bias was being applied. In this state, the increase in current due to the flammable gases was very small. Without a bias, the current returned to the initial value after several hours in dry air.

The flammable gases induced a continuous increase in the current. After the initial rapid increase, the current continuously increased during the introduction of the flammable gas. When the introduction was continued for several hours, the current value did not recover to the level in air as long as bias was applied.

After heating at 250℃, the junctions showed I-V characteristics and gas sensing properties different from those of as-deposited junctions. Figure 6 shows the forward I-V characteristics of the junctions at room temperature in dry air before and after heating at 250℃. With respect to the gas sensing properties, the relationship between the sensitivity and measured temperature changed, as shown in Fig.7.

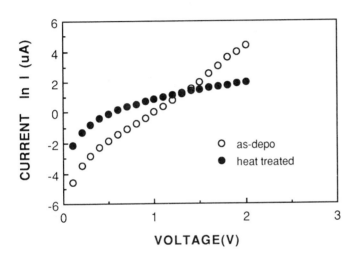

Figure 6. I-V characteristics(forward) at room temperature of as-depo(○) and heat treated(●) junctions.

The temperature dependence of the conductivity of a single semiconductive layer film is shown in Fig.8. In this measurement, conductivity was examined in the direction parallel to the film surface. A large difference in the conductivity between the as-deposited and heat treated film was observed in the ZnO film, and a much smaller difference in the CuO film.

DISCUSSION

From the finding that junctions without patterns show little sensitivity, it is believed that to function as a sensor will

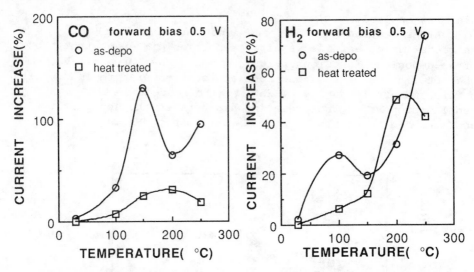

Figure 7. The current increase of as-depo(○) and heat
treated(□) junctions for CO and H₂ -flow.

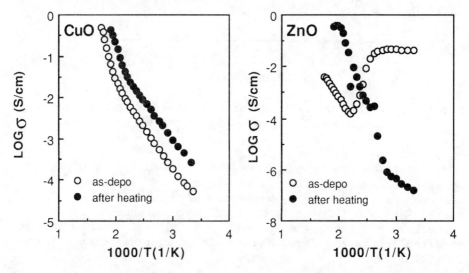

Figure 8. The temperature dependence of the conductivity of the
CuO and ZnO film.

require exposing the interfaces to the atmosphere.

Because the current change occurred only at a forward bias
and the sensitivity changed according to the bias, the sensing

mechanism appears to be different from conventional semiconducting ceramic sensors, which detect a change of conductivity independent of applied bias. A possible sensing mechanism is that CO or H_2 gas adsorb preferentially on CuO, which is known as good adsorbent and catalyst for flammable gases. At the same time, oxidized or hydrogenated species adsorb on ZnO. When a p-n forward bias is applied to the junctions, it is assumed that these adsorbates react with each other and then charge transfer occurs. Because the junctions are made of thin films with resistivities of less than 10^1 Ω, the reactions occur mainly at the interfaces where almost all the biases are applied. As a consequence of these reactions, the carrier concentration, the barrier height, and the density of interface states formed by the adsorbates change, resulting in the increase in forward current. Differences in the energies for the adsorption of CO and H_2 gases are considered to be the cause of the different sensitivity for them.

Differences in the dependence on temperature of the sensing properties between as-deposited and heat treated junctions are believed to result mainly from changes in the ZnO layer. The carrier concentration and activation energy of ZnO change to a large extent. This might be because of the desorption of some adsorbates, surface diffusion of oxygen, or lattice reconstruction.

The continuous increase in the current suggests an increase of carrier concentration and the decrease in the barrier height at the semiconductor interface by the forward bias. The transfer of adsorbates at interfaces (oxygen oxidized and water decomposed species) by the applied forward bias might occur. Not only the transfer of adsorbates at the interfaces but also the diffusion of absorbates from inside films is thought to be occurring. After a continuous increase in the current, the junction showed little sensitivity. This suggests that the adsorbates and the interface states formed by them play important roles in sensing properties. Therefore, the application of higher forward bias, which promotes the transfer of adsorbates, can act negatively with respect to the current increase by flammable gases, particularly at high temperatures. This could explain why at high temperatures the sensitivities (the ratio of current increase) did not increase with the applied forward bias, as shown in Fig.5.

CONCLUSIONS

The interface of a CuO/ZnO thin film was found to function as a flammable gas sensor when the interface is partially exposed to the atmosphere. The application of a p-n forward bias to the interfaces was essential for sensing. Reactions of detecting adsorbates and the change of interface states were suggested to be promoted by applying a forward bias. The sensitivity for CO and H2 gases depended on temperature and the applied bias, and the selectivity of the CO and H2 gases was realized at about 150°C and below 1V bias. After heat treatment of 250°C, the sensing properties changed, largely because of changes in the ZnO film. Utilizing the interfaces of different semiconductive films was found to be a promising novel concept for a chemical sensor.

REFERENCES

1.Tonomura, S., Matsuoka, T., Yamamoto, N., and Tsubomura, H., Metal-Semiconductor junctions for the detection of reducing gases and the mechanism of the electrical response. Nippon-Kagakukai-Shi., 1980,10,1585-90

2.Nakamura, Y., Ikejiri, M., Miyayama, M., Koumoto, K., and Yanagida, H., The current-voltage characteristics of CuO/ZnO Heterojunctions. Nippon-Kagakukai-Shi., 1985,6, 1154-59

3.Nakamura, Y., Tsurutani, T., Miyayama, M., Okada, O., Koumoto, K., and Yanagida, H., The detection of carbon monoxide by the oxide-semiconductor hetero-contacts. Nippon-Kagakukai-Shi., 1987,5, 477-83

4.Miyayama, M., Yatabe, H., Nakamura, Y., and Yanagida, H., Chlorine gas sensing using ZnO/SiC hetero-contact. Yogyo-Kyokai-Shi., 1987, 95, 1145-47

5.Koyama, R., Nakamura, Y., Miyayama, M., and Yanagida, H., Hetero-contact type chlorine sensor. Proc. 1st Japan International SAMPE Symposium., 1989, 375-80

6.Nakamura, Y., Koyama, R., Koumoto, K., and Yanagida, H., Oxidizing gas sensing by SiC/ZnO heterocontact-NOx sensing-. Nippon Seramikkusu Kyokai Gakujutu Ronbunshi.,1991,99, 823-25

7.Ushio, Y., Miyayama, M., and Yanagida, H.,Fabrication of semiconductor open-junction using lithography process and application to chemical sensor. Chemical sensors.,1991,7, Supplement A.,Digest of the 12th Chemical Sensor Symposium., 141-44

8.Ushio, Y., Miyayama, M., and Yanagida, H., Fabrication of thin film CuO/ZnO heterojunction and its humidity sensing properties. submitted to Sensors and Actuators.

Session II

Evaluation

THE QUANTITATIVE EVALUATION OF GRAIN BOUNDARY PHASES BY CRYOGENIC SPECIFIC HEAT MEASUREMENTS

TOYOHIRO HAMASAKI and KOZO ISHIZAKI
Nagaoka University of Technology
Nagaoka, Niigata 940-21, JAPAN

ABSTRACT

The quantity and types (i.e., crystalline or glassy structure) of grain boundary glassy phases can be evaluated by cryogenic specific heat measurements. The specific heat of three types of Si_3N_4 ceramics treated by different processes from the same raw powder and with the same amount of Y_2O_3 and Al_2O_3 additives was measured at different temperatures between 10 and 50 K. The difference between the measured specific heat and the calculated specific heat from the Debye theory confirmed the existence of a glassy phase in the HIPped Si_3N_4. The amount of grain boundary phase should be lower in the Si_3N_4 ceramics than in the surface modified raw powder, which has less surface oxides. This was confirmed by the cryogenic heat capacity measurements. The amount of glassy phase in the heat treated Si_3N_4 ceramics after the same HIP sintering decreased due to crystallization of the grain boundary glassy phase. This was also confirmed by the same measurements. The difference in the temperature dependence of these two samples with reduced grain boundary phases indicates the quantitative difference between the grain boundary glassy phase and the recrystallized oxynitride grain boundary phase.

INTRODUCTION

It is known that the presence of a grain boundary glassy phase in sintered Si_3N_4 has a deleterious effect on mechanical properties, particularly at high temperatures [1,2]. Therefore the quantitative evaluation of the amount of grain boundary glassy phase is of great importance.

Usually the amount of grain boundary glassy phase is determined by destructive metallographic methods, such as a chemical and thermal etching [3–5].

In this study, the grain boundary phase was evaluated from measurements of the specific heat with a low temperature adiabatic calorimeter. The measured specific heat was compared with the calculated specific heat from the Debye theory. The proposed method is non–destructive, and the sensitivity can be extremely high. This method could be very useful for evaluating small changes in the amount of grain boundary phases.

EXPERIMENTAL PROCEDURE

Three types of Si_3N_4 ceramics were investigated in this study. The first (sample A) was sintered from a commercial Si_3N_4 powder (Ube Industry Co., Ltd., UBE–SN–E10, specific surface area 11.1 m^2/g, average particle size 330 nm). The second (sample B) was made from the same powder with modified surfaces by aqueous Soxhlet washing. This modified powder has less surface oxides [12]. Powders of Y_2O_3 (Shin–Etsu Chemical Co., Ltd., 1.3 m^2/g, 900 nm) and Al_2O_3 (Onoda Cement Co., Ltd., 1.7 m^2/g, 900 nm) as sintering aids were added (3 mol% each) to both Si_3N_4 powders. The third type (sample C) was the same sample as A with a post–sintered heat treatment in argon at 1300°C (1573 K) for 10 hours.

The powders were mixed for 24 hours in ethanol by a ball milling machine with Si_3N_4 balls in a plastic vessel. The powder slurries were dried in a vacuum furnace at 50°C (323 K) for 3 hours and crushed. The resultant powder was uniaxially pressed to form pellets 20 mm in diameter and about 5 mm thick, and then cold isostatically pressed at 100 MPa for 60 s. The pellets were coated with BN powder (Shin–Etsu Chemical Co., Ltd.), and encapsulated in a Vycor glass tube after 2 hours degassing at 1000°C (1273 K) in a vacuum of about 1 Pa. The capsules were hot isostatically pressed (HIPped) (Kobelco System 20JT) at 1700°C (1973 K) under 60 MPa pressure for 1 hour. Density measurements by the Archimedean method showed full densification.

To modify the surface of the particles, commercial Si_3N_4 powder was washed with deionized water in a Soxhlet extractor for 100 hours. Experimental details are given elsewhere [11–13].

The specific heat was measured at temperatures from 10 to 50 K using a laboratory designed adiabatic calorimeter. Figure 1 shows a part of the equipment for the specific heat measurements. The specific heat was calculated from the relationship, $\Delta Q/[W \cdot \Delta T]$, where ΔQ, ΔT, and W are the added heat, the temperature change, and the sample weight, respectively. The temperature was measured using a carbon glass resistor.

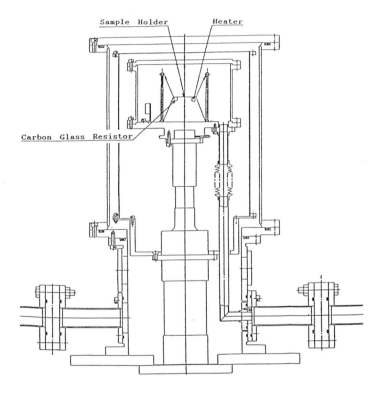

Figure 1. Schematic diagram of a part of the equipment for specific heat measurement. The temperatures from 10 to 50 K were measured by using a carbon glass resistor.

RESULTS AND DISCUSSION

The isochoric specific heat of an electrically insulating material with a perfect crystalline phase at low temperatures decreases in accordance with the Debye T^3 law [8]. At cryogenic temperatures the difference between the isochoric specific heat and the isobaric one can be ignored. The transition to a noncrystalline state is associated with a departure from the Debye T^3 law [7]. The difference between the specific heat of noncrystalline and crystalline phases has been referred to as an abnormal specific heat. It is caused by some extra modes characteristic for the glassy phase [7]. Figure 2 shows the results of the specific heat measurements for sintered Si_3N_4 ceramics A, B, and C. The Si_3N_4 specific heat is a sensitive function of the temperature,

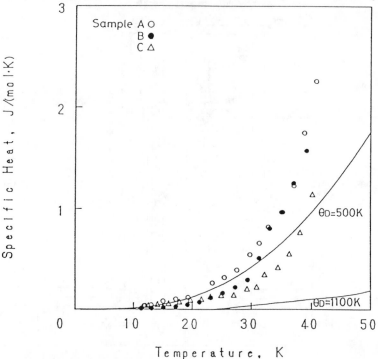

Figure 2. Specific heat of HIPped Si_3N_4 vs. temperature. Open circles and solid circles represent specific heat of the HIPped Si_3N_4 made from the powder (sample A) and from the same powder after 100 hours aqueous washing (sample B), respectively. The open triangle (sample C) indicate the heat treated sample A. The lines correspond to the calculated specific heats at Debye temperatures of 500 K and 1100 K. The amount of glassy phase was decreased by annealing and washing.

particularly, in the low temperature regime. As seen here the calculated lattice heat capacity of θ_D=1100 K is negligible compared with the heat capacity of other crystals e.g., θ_D=500 K, and the measured heat capacities. To evaluate them properly, they were also plotted on a logarithmic scale, as shown in Figure 3. Figure 3 also shows the data from a previous study below 10 K [9].

In the present study, Si_3N_4 ceramics were made using of Y_2O_3 and Al_2O_3 as sintering aids to provide liquid–phase sintering. During densification these additives react with silicon nitride surface oxides to form a liquid phase [10]. Upon cooling the liquid phase remains at the grain boundaries and forms a glassy phase.

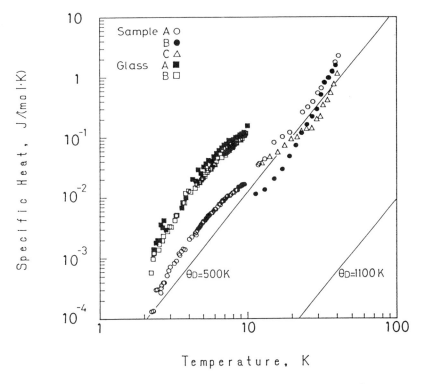

Figure 3. Similar plots of Figure 2 in logarithmic scale. The symbols for the samples are the same as in Figure 2. Solid and open squares represent the specific heats of glasses whose compositions were $8.0Si_3N_445.9Y_2O_320.7Al_2O_325.4SiO_2$ (Glass A) and $15.0Si_3N_442.4Y_2O_319.1Al_2O_323.5SiO_2$ (Glass B) in wt%, respectively. The lines correspond to calculated specific heats from Debye temperatures of 1100 and 500 K.

The presence of a glassy phase in the sintered samples explains why the measured specific heat exceeds the calculated value from the Debye temperature of 1100 K [6] in all temperature ranges, as shown in Figure 3.

As it has been shown, the modification of silicon nitride particle surfaces by aqueous Soxhlet washing leads to dissolution of surface oxide layers. Significant changes in the amount of physi– and/or chemisorbed species (H_2O, NH_3, CO_2, SiO, etc.) were observed [11–13]. Sintering the surface modified powder that has a smaller amount of surface oxides reduces the amount of liquid phase formed and, consequently, there is a smaller amount of glassy phase in the sintered ceramic. Therefore the specific heat of the sample made from the surface modified powder is lower than the specific heat of the sample made from the original powder (Figure 3). The decreased amount of liquid phase produced during the sintering of surface modified Si_3N_4 powders is evidenced also by a slower α–β transition rate compared with unmodified commercial powders [14].

Post–sintering heat treatment of Si_3N_4 ceramics is considered to be a way to improve the mechanical properties and oxidation resistance of liquid–phase sintered materials due to crystallization of the grain boundary glassy phases [4,15]. The HIP sintered sample A was also heat treated, in the present research, to examine the modification of the grain boundary phase. Figure 3 shows the specific heat of this heat treated sample (C in Figure 3).

The specific heat can be represented by:

$$C_p = x C_{1,\,\theta=1100} + y C_{2,\,1100 > \theta > \theta \text{ of glass}} + z C_3 \tag{1}$$

where x, y, and z are respectively the amount of grain lattice, grain boundary crystalline phase, and grain boundary glassy phase, and C_p, C_1, C_2, and C_3 are respectively the total specific heat, the lattice specific heat calculated from the Debye temperature θ of 1100 K, the specific heat of the partially crystallized grain boundary phase (probably θ_D lower than that of Si_3N_4), and the specific heat of the grain boundary glassy phase.

In Figure 3, the specific heat of glass does not decrease in accordance with the Debye T^3 law, and indicates a larger value than the specific heat of the HIPped Si_3N_4. The measured specific heat of the HIPped Si_3N_4 is close to the lattice specific heat with a Debye temperature of around 500 K, although the high temperature heat capacity is close to that of 1100 K as seen in Figure 4 [6]. The specific heat of the heat treated sample (C) is smaller than the sample (B) made from the modified powder at temperatures higher than 25 K, and agrees with the value of the sample

HIPped (A) below 15 K. In this range of low temperatures, C_1 is negligible as can be

seen in

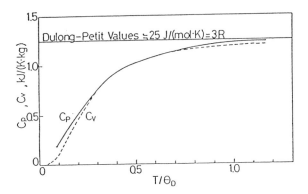

Figure 4. Temperature dependence (T/θ_D) of observed isobaric specific heat (C_P) of sample A and isochoric specific heat (C_V) ca lcula ted from the Debye theory for HIPped Si_3N_4, using the modified Debye temperature of θ_D=1100K.

Figures 2 and 3. Eq. (1) can thus be re-written as:

$$C_P = yC_2 + zC_3 \tag{2}$$

Sample B has fewer oxygen atoms on the grain boundaries. In other words, the

smaller oxygen contents mean lower y and z than for sample A. Similarly, sample C

has less glassy phase than sample A due to crystallization during heat treatment, i.e.

z of C is lower than A, but y of C is higher than A. The temperature dependence of

C_2 follows the Debye T^3 law, but C_3 does not. Samples C and B have different

temperature dependencies due to the different heat capacities C_2 and C_3 of Eq. (2).

The conventional methods for grain boundary quantitative evaluation are

destructive. The present method is non-destructive. It is also sufficiently sensitive to

evaluate small changes in the amounts and kinds of grain boundary phases.

CONCLUSION

Cryogenic heat capacity measurements were conducted for three different Si_3N_4

ceramics. Previously it was not possible to quantify small differences in grain boundary phases. The following was concluded from the present study.

1. The amount and the characteristic of grain boundary phases can be evaluated by specific heat measurements at low temperatures. This method is non–destructive and sensitive enough to evaluate small changes in the amounts and kinds (i.e., crystalline or glassy phase) of grain boundary phases.

2. The amount of grain boundary phases in the HIPped Si_3N_4 made from surface modified powder (aqueous washed) is smaller than in the HIPped Si_3N_4 made from commercial powder.

3. The amount of glassy phase in the post sintered heat treated Si_3N_4 decreased due to crystallization of the grain boundary glassy phase.

4. The difference in the temperature dependence of the heat treated sample and the HIPped one compared with the surface modified powder, indicates that during heat treatment some part of the grain boundary glassy phase crystallizes.

ACKNOWLEDGMENT

The authors wish to express their gratitude to Kobe Steel to allow to HIP their samples. This work was supported in part by a Grant–Aid for the Scientific Research from the Ministry of Education in Japan.

REFERENCES

1. Hirosaki, N., Okada, A., and Mitomo, M., Effect of oxide addition on the sintering and high–temperature strength of Si_3N_4 containing Y_2O_3. J. Mater. Sci., 1990, **25**, 1872–76.

2. Zeng, J., Yamada, O., Tanaka, I., and Miyamoto, Y., Hot isostatic pressing and high–temperature strength of silicon nitride–silica ceramics. J. Am. Ceram. Sci., 1990, **73**, 1095–97.

3. Lee, W.E., and Hilmas, G.E., Microstructural changes in β–silicon nitride grains upon crystallizing the grain–boundary glass. J. Am. Ceram. Sci., 1989, **72**, 1931–37.

4. Cinibulk, M.K., and Thomas, G., Grain–boundary–phase crystallization and

strength of silicon nitride sintered with a YSiAlON glass. J. Am. Ceram. Sci., 1990, **73**, 1606–12.

5. Babini, G.N., Bellosi, A., and Vincenzini, P., Factors influencing structural evolution in the oxide of hot-pressed Si_3N_4-Y_2O_3-SiO_2 materials. J. Mater. Sci., 1984, **19**, 3487–97.

6. Watari, K., Seki, Y., and Ishizaki, K., Temperature dependence of thermal coefficients for HIPped silicon nitride. J. Ceram. Soc. Jpn., 1989, **97**, 174–81, and J. Ceram. Soc. Jpn. International, 1989, **97**, 170–78.

7. Zeller, R.C., and Pohl R.O., Thermal conductivity and specific heat of noncrystalline solids. Phys. Rev. B, 1971, **4**, 2029–41.

8. For example, Rosenberg, H.M., Low temperature solid state physics, Oxford Univ. Press, London, 1963, pp. 7.

9. Watari, K., Ishizaki, K., and Mori, K., Evaluation of glassy phase in sintered silicon nitride by low-temperature specific heat measurements. J. Am. Ceram. Soc., 1991, **74**, 244–46.

10. Loehman, R.E., Oxynitride glasses. J. Non-Cryst. Solids, 1980, **42**, 433–46.

11. Kawamoto, M., Ishizaki, K., and Ishizaki, C., Characterization of silicon nitride powder by a temperature programmed desorption study, In Euro-Ceramics, Properties of Ceramics, ed. G. de With, R. A. Terpstra and R. Metselaar, Elsevier Applied Science, London and New York, 1989, **2**, pp.120–24.

12. Kawamoto, M., Ishizaki, C., and Ishizaki, K., Fluidity-increasing behavior of silicon nitride powder by aqueous washing. J. Mater. Sci. Letters, 1991, **10**,, 279–81

13. Ishizaki, K., Kawamoto, M., and Ishizaki, C., Improving flow behavior of silicon nitride powder by surface modification, in Surface Modification Technology IV, ed., T. S. Sudarshan, D. G. Bhat and M. Jeandin, The Minerals, Metals and Materials Society, Pennsylvania, USA, 1991, pp.655–63.

14. Saito, N., Ishizaki, K., and Kuzjukevics, A., The production of porous Si_3N_4 with 100% β phase. in Pressure Effects on Materials Processing and Design, ed. K. Ishizaki et al., Vol. 251, Materials Research Society, Pittsburgh, Pennsylvania, USA, 1992, pp.149–154.

15. Falk, L.K.L. and Dulop, G.L., Crystallization of the glassy phase in an Si_3N_4 material by post-sintering heat treatment. J. Mater. Sci., 1987, **22**, 4369–76.

GRAIN BOUNDARY ANALYSIS OF SILICON NITRIDE MATERIALS

Tsuneo MISHIMA, Rieko TANAKA, Shoji KOHSAKA and Kazunori KOGA
CENTRAL RESEARCH LABORATORY, KYOCERA CORPORATION
KOKUBU CITY, KAGOSHIMA 899-43, JAPAN

ABSTRACT

A method to determine the composition of the grain boundary phases in sintered silicon nitride was derived using an AEM (Analytical Electron Micriscope) with an UTW-EDX (Ultra-Thin Window Energy Dispersive X-ray spectrometer). The composition of the specimen was determined by the peak ratio of the EDX spectra for rare earth, silicon, nitrogen, and oxygen ions.

The absolute error of measurement using an UTW-EDX and the deviation from the results of chemical analysis were confirmed to be within $\pm 1\%$ for all elements. The dispersion of measurement was confirmed to be $\pm 2.3\%$.

INTRODUCTION

Silicon nitride has been considered a promising candidate for high-temperature heat engine applications. The mechanical, chemical, and thermal properties of silicon nitride are mainly determined by the composition and state of the grain boundary phases[1]. In order to develop high-performance silicon nitride with high strength at high temperatures, high fracture toughness, and high corrosion resistance, it is indispensable to develop a method for compositional control and analysis of the grain boundary phases.

The mechanical properties of silicon nitride are usually tailored by the selection of the sintering additives, such as Al_2O_3, Y_2O_3, MgO, ZrO_2 etc. and by heat treating to promote crystallization. Varying the sintering additives changes the composition of the grain boundary phases in the silicon nitride. The composition of these phases also varies depending on the degree of reaction between silicon nitride and silicon dioxide on the surface of the silicon nitride powder. Usually the reaction is accompanied by a compositional change due to vaporization of reaction products such as silicon monoxide and nitrogen gas. Therefore, otaining heat resistant silicon nitride requires not only selecting suitable sin-

tering additives but also optimizing the sintering processes[2] and analyzing the grain boundaries.

Determination of the accurate composition of the grain boundary
phases of silicon nitride was attempted in the present study using a TEM
(Transmission Electron Microscope). TEM is one of the most effective ways
to obtain information about grain boundary phases. The microstructure of
the grain boundary phase plays an important role in the mechanical properties of a material. The degree of crystallization of the grain boundary
can be estimated by using XRD (X-Ray Diffraction). However, XRD can not
distinguish whether or not a grain boundary phase is fully crystalline. In
contrast, TEM can detect a small amount of amorphous film between silicon
nitride crystals or between silicon nitride crystals and a crystalline
grain boundary phase.

Several results have been reported for the compositional analysis of
the grain boundary phases of silicon nitride[3,4]. These used the Cliff-
Lorimer method to obtain quantitative values for the composition. However,
there still remains ambiguity in the data, because correcting for the
specimen thickness is difficult.

This paper discusses the application of the DXA method by Nemoto and
Horita to the compositional analysis of the grain boundary phases of silicon nitride.

EXPERIMENTAL

Sample preparation

Compositions in the silicon nitride-rare earth oxide-silica ternary system
were selected for simplicity and because of the amount of works on heat
engine applications that has been done.

Three rare earth-silicon oxynitride glasses with different compositions were prepared. Uniform glass specimens were prepared by melting the
mixtures of powder composition listed in Table 1. Powder compacts of each
composition were heated up to 1800°C in a pressurized nitrogen atmosphere
(0.98MPa) to prevent the decomposition of Si_3N_4 at high temperatures. The
glasses showed no discernible crystalline diffraction peak in the XRD
measurement and no crystallites by TEM observation.

TABLE 1. Compositions of the glasses by chemical analysis (at%)

Specimen	Rare Earth	Silicon	Oxygen	Nitrogen
1	18.2	19.3	52.5	10.0
2	15.6	20.8	57.5	6.0
3	20.8	21.0	51.4	10.7

Electron-transparent specimens were prepared by slicing, mechanical
polishing, and Ar milling.

Theory of quantification by AEM[5-9]

According to the theory of Cliff-Lorimer, the intensity ratios of the
characteristic X-rays of the elements are proportional to the ratio of the

amounts of the elements at the coherent region of the electron beam path.

$$(C_1/C_2)=k(I_1/I_2) \tag{1}$$

where,

       C_i:mass fraction of the element i
       I_i:characteristic X-ray peak intensity from element i
       k ;k-factor, coefficient of concentration and intensity ratios

However, to obtain more precise compositional data, a correction of the observed intensity for the absorption along the characteristic X-ray path is required. For this reason the Cliff-Lorimer method is modified in the present analysis because of the high absorption of the characteristic X-rays from light elements. Using the absorption correction factor,CF, derived by Goldstein, the intensity ratio of characteristic X-rays can be converted to the concentration ratio, as given by equation (2), which is rewritten from Eq.(1),

$$(C_1/C_2)=CF*k_0(I_1/I_2) \tag{2}$$

For a thin specimen, the correction factor, CF between the concentration and the characteristic X-ray intensity can be expressed by:

$$CF= \frac{(\mu/\varsigma)_A(1-\exp[-\{(\mu/\varsigma)_B \cosec\ a(\varsigma*t)\}]}{(\mu/\varsigma)_B(1-\exp[-\{(\mu/\rho)_A \cosec\ a(\varsigma*t)\}]} \tag{3}$$

where,

      μ/ς ;mass absorption coefficient
      a ;take off-angle
      $\varsigma*t$;density*thickness.

Alhough the absorption correction factor CF is given as a function of specimen thickness $\varsigma*t$ and the mass absorption coefficients μ/ς, it is difficult to determine the specimen thickness. However, the DXA (Differential X-ray Absorption) method proposed by Nemoto and Horita[7-9] is useful to determine the CF by calculation without measuring the specimen thickness. Also the method was improved specifically to obtain exact data for the spectra of light elements.

Determination of the composition
The measurement of the intensity of characteristic X-rays was carried out using JEM2000FX type AEM equipped with Tracor Northern Series II UTW-EDX. Determination of the composition of the grain boundary phase in sintered silicon nitride was carried out using silicon nitride-silica-rare earth oxide glasses as standard samples. It includes the determination of the amount of RE(rare earth element) and Si as well as light elements such as oxygen and nitrogen. The AEM was operated at 200KV and with a specimen tilting angle of 40°. Specimen No.1 was used to determine k-factors. The results of the AEM analyses were the averages of 8 or 10 measurements.

RESULTS AND DISCUSSION

Figure 1 shows a transmission electron micrograph and X-ray spectrum for the grain boundary phase in a sintered silicon nitride. Dark phases are the grain boundary phase. This includes heavy rare-earth elements, silicon, oxygen, and nitrogen. The size of the grain boundary phase is 0.5 to 1.0 um.

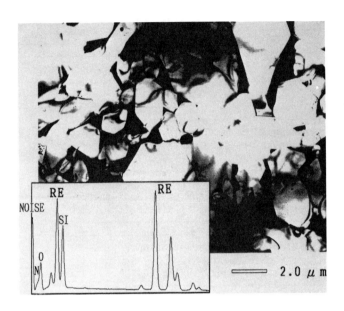

Figure 1. TEM image and EDX spectrum for the grain boundaries of silicon nitride sintered with a rare-earth additive.

The process to determine the k-factor for each composition and thickness of specimen is illustrated in Fig.2 by a flow chart. Figure 3 shows an X-ray spectrum obtained for a rare-earth silicon-oxynitride glass. With increasing specimen thickness, the ratio of X-ray intensities I_{RE-M}/I_{RE-L} decreases. The change of the peak ratio is due to the X-ray absorption in the specimen. The calibration of the specimen thickness was carried out based on the DXA method using two characteristic peaks of the rare earth ion.The thickness and the CF can be determined using the characteristic X-ray intensity ratio between the rare-earth M line and the rare-earth L line.

The mass absorption coefficients for all constituent elements are actually determined using points with different thicknesses in a specimen. The k-factors for each composition and thickness of a specimen are determined as illustrated in the flow chart in Fig.2. In Eqs.(1)-(3) other parameters than $\rho*t$ can be determined by the result from the measurement of standard materials such as rare earth oxides, rare earth silicates, and standard glasses. The calculated values of mass absorption coefficients of i element for the j characteristic X-ray, $(\mu/\rho)_i^j$, are used to calculate the mass absorption coefficient for the specimen of composition C_i, as:

$$(\mu/\rho)^{j}_{spec}=sum_i C_i (\mu/\rho)^{j}_i \qquad\qquad (4)$$

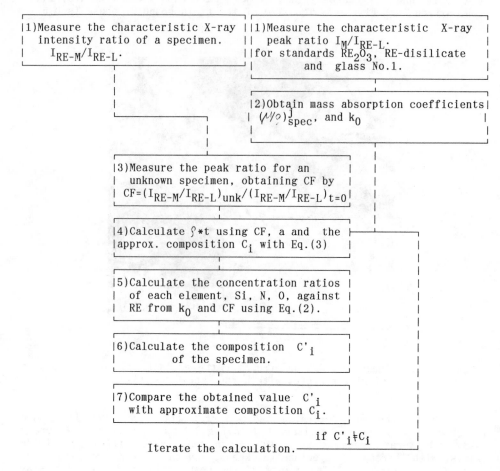

Figure 2. Calculation scheme for the composition from characteristic X-ray intensities.

The intensity ratios of characteristic X-rays obtained using the absorption corrections are plotted in Fig.4 against the intensity of the rare-earth L line (the intensity of rare earth-L line was used to represent sample thickness). If the absorption correction was properly done, the ratios should be constant. However, as shown in Fig.4, the corrected values using the mass absorption coefficients of Heinrich are slightly over-corrected. In this study, the calculations of absorption coefficients were optimized until corrected values becomes constant.

The compositions of the glasses determined by the UTW-EDX method are compared in Table 2 with the chemical analysis data. The absolute error of measurement using UTW-EDX, i.e the deviation from the results of chemical analysis, was confirmed to be within ±1% for all the elements. The dispersion of measurement shown in Table 2 was confirmed to be $\pm 2.3\%$.

TABLE 2. Comparison between the results of UTW-EDX and chemical analysis
(at%)

Specimen	Method	Rare earth	Silicon	Oxygen	Nitrogen
2	chemical	15.6	20.8	57.5	6.0
	UTW-EDX	15.1$\pm$1.3	21.6$\pm$0.5	56.5$\pm$1.8	6.8$\pm$1.4
3	chemical	17.0	21.0	51.4	10.7
	UTW-EDX	16.7$\pm$1.5	21.2$\pm$0.7	52.4$\pm$1.7	9.7$\pm$2.3

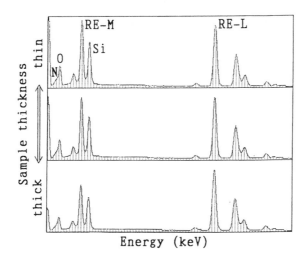

Figure 3. Characteristic X-ray spectra for different specimen thichnesses.

Although a probe diameter of 100 nm was used, the minimum area of analysis needed to be about 300 nm to carry out accurate analysis in considering the X-ray absorption path length. The method developed here was applied to determine the composition of the grain boundary phase of a sintered silicon nitride material. The initial composition of the mixture is indicated by an open circle in Fig.5. After sintering, if the total composition does not change, the composition of the grain boundary phase among silicon nitride crystals should be on the line connecting silicon nitride and the rare earth disilicate. However, the analysis result shows that the composition of the grain boundary phase moved to the point desig-nated by ●, where a low temperature eutectic composition of end components of the tie-line exists. This result suggests that the composition of the grain boundary phase is not determined by the starting powder composition but by the lowest eutectic composition in the compatibility triangle.

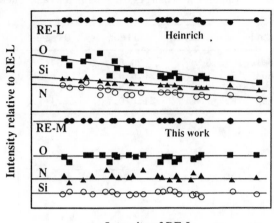

Figure 4. Variation of intensity ratio I_{RE-M}/I_{RE-L} as a function of I_{RE-L}.

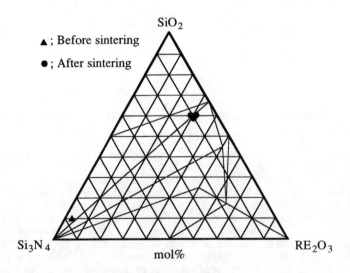

Figure 5. Compositional change of the grain boundary phases during the sintering of silicon nitride.

CONCLUSIONS

Two rare-earth silicon oxynitride glasses with different composition were prepared for analysis. The accuracy of the present quantitative analysis was studied by comparing the results of the AEM analysis with those of chemical analysis. The absolute and relative dispersions of the AEM analysis were confirmed to be within 1% and 20%, respectively, including light elements.

The application of the DXA method of Nemoto and Horita and the modification of mass absorption coefficients by the AEM method provide accurate quantitative analysis of the nitrogen and oxygen at the grain boundary phases in sintered silicon nitride materials.

ACKNOWLEDGEMENT

The authors wish to thank Prof.s Nemoto and Horita of Kyusyu University for their instructions and discussions on the DXA method.

References;
1)K.Koga;Bulletin of the Ceramic Society of Japan,25[2](1989)107-11(1990).
2)S.Kohsaka and K.Koga;Proceedings of Science of Engineering Ceramics,ed.
  by Kimura and Niihara,The Ceramic Society of Japan,Tokyo Japan(1991)
  589-594.
3)Y.Bando,M.Mitomo and Y.Kitami;J.Electron Microsc.vol.35,No.4(1986)
  371-77.
4)M.H.Lewis and A.R.Bhatt;J.Mater.Sci.,15(1980)103-13.
5)P.L.Morris,M.D.Ball and P.J.Statham;Electron microscopy and Analysis
  1979,T.Mulvey ed. Inst.Phys.Conf.Ser.No.52C(1980)413.
6)Practical Analytical Electron Microscopy in Materials Science,David B.
  Williams ed.(1984) verlag chemie international.
7)Z.Horita;Doctoral Theses, Dept. of Materials Science and Engineering
  Kyushu University(1991).
8)Z.Horita,T.Sano and M.Nemoto;Bulletin of the Japan Institute of Metals,
  vol.28,No.9,742(1989).
9)Z.Horita,K.Ichiki,T.Sano and M.Nemoto;Philosophical Magazine A,vol.59,
  No.5,939-52(1989).

EVALUATION OF GRAIN BOUNDARY PHASES OF ß-SIALON BY CRYOGENIC SPECIFIC HEAT MEASUREMENTS

TOYOKAZU KURUSHIMA[*], KOZO ISHIZAKI[**] AND TOYOHIRO HAMASAKI[**]
*INAX Corporation, Tokoname, Aichi 479, JAPAN
**Nagaoka University of Technology, Nagaoka, Niigata 940-21, JAPAN

ABSTRACT

Fully densified ß-SIALON ceramics ($Si_6Al_2O_2N_8$), were prepared by a normal sintering and a post-sintering HIP (hot isostatic pressing) treatment. After exposure to corrosion by molten copper oxide (CuO) at 1200℃ in air for 1 hour, both samples were evaluated by measuring their specific heat at cryogenic temperatures. Due to the fast cooling rate during sintering, the HIPed sample contained more grain boundary glassy phases than the normally sintered one. This was evaluated by a newly developed copper oxide etching method. Cryogenic specific heat is a powerful and a sensitive property for evaluating the grain boundary glassy phases of ß-SIALON ceramics. The copper oxide etching method used was found to be an effective method for observing ceramic structures with siliceous grain boundary glassy phases.

INTRODUCTION

ß-SIALON ceramics are considered one of the best refractory materials for containing molten metals such as steel [1,2], but they are easily corroded by molten copper oxide [3]. However, they are not corroded to any extent by steel slag. It is possible that their corrosion by copper oxide could be caused by the presence of grain boundary glassy phases.

Watari et al. have proposed that the abnormal specific heat at cryogenic temperatures of silicon nitride ceramics could be used to evaluate glassy phases [4]. Hamasaki et al. showed that this is also an extremely sensitive method for evaluating the amount of grain boundary glassy phases in silicon nitride ceramics [5].

This study uses cryogenic specific heats to compare the relative amounts of glassy phases in HIPed and normally sintered ß-SIALON ceramics ($Si_6Al_2O_2N_8$). The relation of the amount of grain boundary glassy phases present in ß-SIALON ceramics to their rate of corrosion by molten copper oxide is discussed.

MATERIALS AND EXPERIMENTAL METHODS

TABLE 1 shows the grain sizes and chemical compositions of the starting materials for ß-SIALON ceramics. The raw materials were pressed uniaxially at 20 MPa, and CIPed (cold isostatic pressing) at 100 MPa.

TABLE 1
Grain size and chemical composition of
starting materials for ß-SIALON

Material	Average Grain Size	Composition
ß-SIALON ($Si_4Al_2O_2N_6$)	0.6 μm	99 wt%
Y_2O_3	3 μm	1 wt%

The sintering schedules are shown in **Figure 1**. Normal sintering is indicated by a solid line. The HIPed samples were first sintered normally following the solid line in this figure, and then HIPed at 100 MPa of nitrogen gas following the dashed line. Using a diamond cutter, the samples were cut into rectangular shapes about 10 mm X 10 mm X 7 mm. The density was calculated by Archimedes' principle after weighing in water and air.

Figure 2 is a schematic diagram of the corrosion test. The samples were heated in molten copper oxide at 1200℃ for 1 hour. The alumina lid of the heating vessel was kept slightly open to maintain the oxygen partial pressure at ambient pressure. Samples before and after the corrosion test were studied by scanning electron microscopy (SEM).

The specific heats of the ß-SIALON samples were measured at temperatures between 10K and 40K using equipment developed by Hamasaki et al. [5].

RESULTS

The normally sintered and the HIPed ß-SIALON ceramics were fully densified to the densities of 99.5% and 100% respectively.

Figure 3 shows photographs of ß-SIALON samples before and after the corrosion test. A marked reduction in dimensions occurred, particularly for the HIPed sample.

Figure 4 shows the specific heats of the normally sintered and the HIPed ß-SIALON samples at temperatures between 10K and 40K. For reference, the solid line indicates the specific heat of silicon nitride ceramics calculated from the Debye theory for temperatures of ·=1100 K [4]. The difference between the specific heats of the HIPed ß-SIALON and the normally sintered sample is mainly due to the amount of grain boundary glassy phases present. One can observe that the HIPed sample contains more grain boundary glassy phases.

Figure 5 shows the SEM micrographs of polished surfaces of the corrosion tested ß-SIALON samples. Pores are located along the grain boundaries. More pores are present in the HIPed sample than in the normally sintered one.

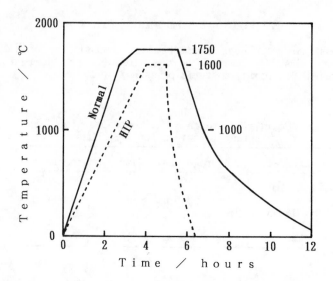

Figure 1. Sintering schedules of ß-SIALON ceramics. In the normal sintering case (the solid line), samples were heated to 1600℃ at a rate of 10℃/min., then to 1750℃ at 3℃/min., held at this temperature for 2 hours, cooled down to 1000℃ at a rate of 10℃/min., and cooled to room temperature slowly in a furnace. In the HIP sintering case, samples were normally sintered, shown by the solid line, and then were HIP treated, shown by the dashed line. Samples were heated to 1600℃ at a rate of 7℃/min., held at this temperature for 1 hour, and cooled to room temperature rapidly at a rate of about 20℃/min. in the HIPing furnace.

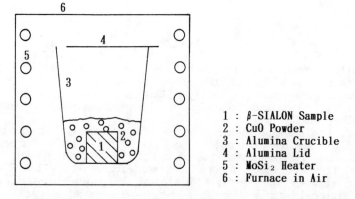

1 : ß-SIALON Sample
2 : CuO Powder
3 : Alumina Crucible
4 : Alumina Lid
5 : MoSi$_2$ Heater
6 : Furnace in Air

Figure 2. Schematic diagram of the corrosion test. A ß-SIALON sample plus CuO powder were heated in an alumina crucible at 1200℃ in air for 1 hour. The alumina lid was kept slightly open to maintain the oxygen partial pressure at ambient pressure.

DISCUSSION

Cryogenic Specific Heat

At temperatures between 10K and 40K, HIPed ß-SIALON ceramics have higher specific heats than those that are sintered normally, as shown in **Figure 4**. The difference between the specific heats of the HIPed ß-SIALON and the normally sintered sample is mainly due to the heat capacity of the glassy phase in the grain boundaries. A HIPed sample that contains a larger amount of grain boundary glassy phases has a higher specific heat at cryogenic temperatures than one sintered normally [4].

The HIPed sample is also more severely corroded by molten copper oxide than a normally sintered one, because it contains more grain boundary glassy phases. The difference between the quantity of glassy phases in the two samples can be very small, but is clearly detected by measuring specific heats. Measurement of the specific heat at cryogenic tempertures is a powerful and a sensitive method for evaluating the grain boundary glassy phases in ß-SIALON ceramics.

Etching by Copper Oxide

Corrosion tests indicate that samples of HIPed ß-SIALON ceramics have more pores than normally sintered samples, as shown in **Figure 5**. The pores are located along the grain boundaries. The HIPed sample could have more pores because it contains more grain boundary glassy phases, as can be seen in **Figure 4**, and is more corroded by molten copper oxide, as shown in **Figure 3**.

ß-SIALON ceramics are readily etched by molten copper oxide because their grain boundary glassy phases are corroded selectively by the copper oxide. To observe the structure of ceramics, they are usually etched using HF. Because HF etching is difficult with ß-SIALON or Si_3N_4 ceramics, their fracture surfaces are often observed to examine their structures [6]. A plasma etching method is also used to observe the microstructure of Si_3N_4 [7]. The copper oxide etching method described here is relatively easy to use because the ceramics simply are heated with copper oxide at 1200℃ in air. This is an effective way to observe the structure of ceramics that have siliceous grain boundary phases.

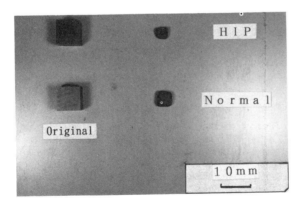

Figure 3. Photograph of ß-SIALON samples before and after the corrosion tests. The HIPed sample was severely affected. The scale marker indicates 10 mm.

Influence of HIPping

Despite its higher density, the HIPed ß-SIALON sample was more severely corroded by molten copper oxide because it contains a larger amount of grain boundary glassy phases. The HIPed sample was cooled to room temperature rapidly as shown in **Figure 1**. Due to the rapid cooling rate, it had more grain boundary glassy phases and was more severely corroded by molten copper oxide than the sample sintered normally. The HIPed samples were not always resistant to corrosion by molten copper oxide if they contained more grain boundary glassy phases.

This slight difference in heat treatment was sufficient to be detected by measuring the cryogenic specific heat. This is indicative of the power and sensitivity of this method for evaluating grain boundary glassy phases.

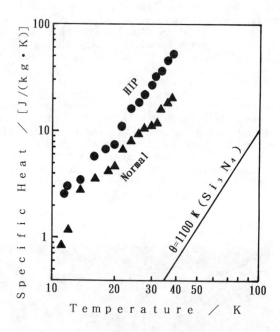

Figure 4. The circles and triangles indicate the specific heats of the HIPed ß-SIALON and normally sintered samples as a function of the temperature, respectively. For reference, the solid line indicates the specific heat of silicon nitride ceramics calculated from Debye theory for temperatures of Θ=1100K [4]. The difference between the specific heats of the HIPed ß-SIALON and the normally sintered sample is mainly due to the amount of grain boundary glassy phases present. One can observe that the HIPed sample contains a larger amount of grain boundary glassy phases.

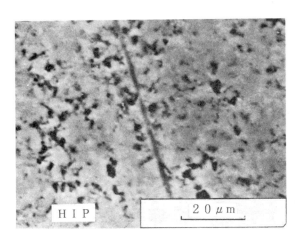

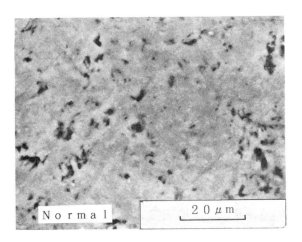

Figure 5. SEM micrographs of the polished surfaces of the corrosion tested ß-SIALON samples. The pores are located along the grain boundaries. There are more pores in the HIPed sample than in the normally sintered sample. The grain boundary siliceous phases in ß-SIALON ceramics can be readily located using copper oxide etching. The scale markers indicate 20 μm.

CONCLUSIONS

1. Measurement of the specific heat at cryogenic temperatures is a powerful and a sensitive tool for evaluating grain boundary glassy phases in ß-SIALON ceramics.

2. Molten copper oxide etches grain boundary glassy phases selectively. This etching method is an effective way to observe ceramic structures that have siliceous grain boundary glassy phases.

3. Due to the fast cooling rate, the HIPed sample contained more grain boundary glassy phases than the normally sintered one. This was observed qualitatively by using a copper oxide etching method. It can be quantitatively evaluated by measuring the cryogenic specific heat.

4. HIPed fully densified samples that contain a large amount of grain boundary glassy phases do not always resist corrosion by molten CuO.

REFERENCES

1. Terao,K., Suzuki,T. and Arahori,T., Erosion resistance of sialon to stainless steel. J.Ceram.Soc.Jpn., 1986, 94, 111-15.

2. Trigg,M.B., Ellson,D.B. and Sinclair,W., Corrosion of selected non oxide ceramics in liquid steel. Br.Ceram.Trans., 1988, 87, 153-57.

3. Kurushima,T. and Ishizaki,K., Reactions of copper and copper oxides with nitride ceramics (AlN,SIALON,Si$_3$N$_4$) and oxide additives (Al$_2$O$_3$,Y$_2$O$_3$,SiO$_2$,MgO). J.Ceram.Soc.Jpn., 1992, 100, (in press).

4. Watari,K., Seki,Y. and Ishizaki,K., Temperature dependence of thermal coefficients for HIPped silicon nitride. J.Ceram.Soc.Jpn., 1989, 97, 174-81, and J.Ceram.Soc.Jpn.International, 1989, 97, 170-78.

5. Hamasaki,T. and Ishizaki,K., Quantitative evaluation of grain boundary phase by cryogenic specific heat measurements. in this conference, 1992.

6. Lis,J., Majorowski,S., Puszynski,J.A. and Hlavacek,V., Dence ß- and α/ß-SiAlON materials by pressureless sintering of combustion-synthesized powders. Am.Ceram.Soc.Bul., 1991, 70, 1658-64.

7. Bocker,W. and Hamminger,R., Advancements in sintering of covalent high-performance ceramics. Interceram, 1991, 40, 520-25.

THE DEPENDENCE OF THE INTERNAL FRICTION AND THE MECHANICAL PROPERTIES OF CERAMICS ON THEIR GRAIN BOUNDARY STRUCTURES

KEN'ICHI MATSUSHITA
Institute of Scientific and Industrial Research, Osaka Univ.
8-1 Mihogaoka, Ibaraki, Osaka 567, JAPAN

ABSTRACT

Measurements were made on internal friction, shear modulus, and mechanical properties of alumina based ceramics from room temperature to 1400°C. The internal friction at high temperatures was strongly dependent on the grain boundary characteristics, such as grain boundary softening, sliding, and the diffusion of ions in the constituents of the grain boundaries. Creep rate and strength at elevated temperatures were also dependent on grain boundary characteristics. Grain boundary structures could be evaluated by using internal friction analysis. Alumina ceramics with improved flexural strength and creep resistance at elevated temperatures were developed by the reduction of ion vacancy concentration in the grain boundary constituents. The fracture toughness and creep resistance were also improved markedly by dispersion of double oxide particles at the grain boundaries.

INTRODUCTION

A variety of oxide, nitride, and carbide ceramics are of industrial importance. However, their mechanical properties vary widely because they are strongly dependent both on the bulk properties and on grain boundary characteristics [1,2,3]. Therefore, the grain boundary characteristics are very important and must be controlled in order to improve the mechanical strength of the ceramics.

It is known that the increment of internal friction in materials corresponds to the movement of atoms, ions, and dislocations due to applied stress and to viscous flow at the grain boundary[4,5]. The measurement of internal friction is superior both to microstructural observation by TEM and SEM and to other mechanical properties tests for detecting microstructure changes and the degradation of mechanical property due to the diffusion of ions in the bulk and in grain boundaries and to grain boundary sliding. Furthermore, internal friction measurements are particularly advantageous for obtaining high temperature data easily and continuously.

In this work, the grain boundary structure of alumina ceramics was investigated using internal friction and shear modulus measurements. The ultimate objective of this work is to improve the grain boundary properties and thereby the mechanical properties at elevated temperatures.

MATERIALS AND METHODS

Materials

Specimens were made as follows; alumina and additives were mixed by ball-milling in ethanol for 10 hours and compacted by uniaxial pressing at 20 MPa pressure to form a 3x10x100 mm bar. The compositions of the mixed powders are shown in Tables 1 and 2. The compacted bar were sintered at 1600°C for 4 hours. The sintered alumina ceramics were polished with a diamond wheel to obtain the specimens for the internal friction and shear modulus measurements. The specimens for the 4-point bend test were cut from the same sintered body. Their surfaces were polished with a 400 grit diamond wheel and the edges of the tensile surface were beveled. The specimen size was 3x4x40 mm.

Table 1. Compositions (mol%) of the specimens used for flexural fracture strength, and internal friction measurements

Specimen	SiO_2	MgO	Ta_2O_5	CaO	Al_2O_3
A		4.9			95.1
B				3.6	96.4
C	5.1	2.0			92.9
D	0.6			2.9	96.5
E	3.0	0.33	0.67		96.0

Table 2. Compositions (mol%) of the specimens used for the creep test and the internal friction measurement

specimen	Y_2O_3	Al_2O_3
A	0	100
0.5Y	0.5	99.5
1Y	1	99
2Y	2	98
5Y	5	95
10Y	10	90

Microstructure observation

Microstructures of the specimens were observed by TEM and SEM. X-ray diffraction was used to identify the matrix and the precipitates in the alumina ceramics.

Internal friction and shear modulus measurements

Internal friction and shear modulus were measured by the torsion pendulum method with a frequency range of 1 to 30 Hz. Internal friction (Q^{-1}) and shear modulus (G) were calculated from the free decay curve and from the resonant frequency of the vibration of the specimen, respectively, as follows:

$$Q^{-1} = \Delta/\pi \qquad\qquad (1)$$

$$G = 3Kf^2L/(1-0.63b/h)hb^3 \qquad\qquad (2)$$

where D is the logarithmic decrement of vibration amplitude per one cycle; f is the resonant frequency of the pendulum; b, h, and L are the thickness, width, and length of the specimen, respectively; and K is an apparatus constant related to the moment of inertia. Internal friction and shear modulus were measured up to 1400°C in an Ar atmosphere with a heating and cooling rate of 5°C/min.

Four-point bending test and creep test

The flexural strength was measured up to 1100°C by using the four-point bending test with a crosshead speed of 0.5 mm/min which corresponds to a surface strain rate of 1.5×10^{-4} /s. The inner and outer span were 10 and 30 mm, respectively. The creep test was carried out at 1200°C to 1400°C under a stress of 30 to 150 MPa in air.

RESULTS

Internal friction and the degradation of strength

Figure 1 shows the temperature dependence of the internal friction and shear modulus in specimens A and B. With increasing temperature, the shear modulus of specimens A and B decreased gradually up to 1200°C. The internal friction in specimen B was nearly 1.5×10^{-3}, but specimen A had a small peak in the vicinity of 947°C. On the other hand, specimen C had an internal friction peak at 907°C and its internal friction markedly increased with increasing temperature above 1080°C, reaching 38×10^{-3} at 1200°C as shown in Fig. 3. The shear modulus of specimen C decreased linearly up to 830°C. Its temperature coefficient increased in the internal friction peak temperature range and then the coefficients increased further above 1080°C. The internal friction peak also appears at 807°C for specimen D, and the internal friction at 1200°C was about 55×10^{-3}. The shear modulus in specimen D decreased linearly up to 730°C largely in the temperature range of the internal friction peak. The internal friction in specimen E is nearly constant up to 930°C, increasing gradually above 930°C but without making a peak, as shown in Fig.2.

X-ray diffraction analysis confirms that the double oxide compounds, $MgAl_2O_3$, $Al_6Si_2O_{13}$, and $CaAl_{12}O_{19}$ were formed in specimen A, in specimens C and E, and in specimens B and D, respectively.

Figure 3 represents the relation between internal friction at 1100°C and the difference between the flexural strength at room temperature and that at 1100°C in specimens A, B, C, D, and E. There is a good relation between the degradation of flexural strength and internal friction. The degradation in specimen A and in its strength at elevated temperatures was the smallest.

Internal friction and the steady state creep rate

Figure 4 shows the temperature dependence of internal friction in specimens containing yttria. No internal friction peak was observed in the internal friction temperature curves of all these specimens. With increasing temperature, internal friction in pure alumina ceramics (AKP-30, Sumitomo kagaku Co.) was nearly constant up to 1130°C and then increased above 1175°C. The internal friction above 1230°C in specimen 1Y was the lowest in this study. The internal friction at 1300°C and 1400°C of specimens containing yttria is represented in Figure 5. With increasing yttria content, the internal friction at 1300°C and 1400°C decreased rapidly up to 1 mol% yttria and then increased gradually. It is noteworthy that the internal friction reaches a minimum value at 1 mol% yttria.

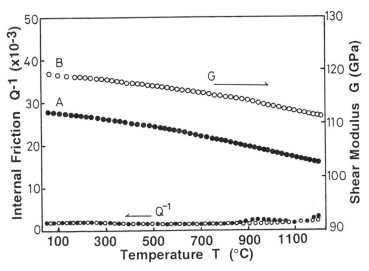

Figure 1.   Temperature dependence of the internal friction and
the shear modulus in specimens A and B.
Solid circles: A   Open circles: B

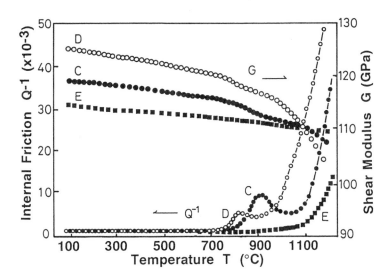

Figure 2.   Temperature dependence of the internal friction and the
shear modulus in specimens C, D, and E.
Solid circles: C   Open circles: D and solid squares: E

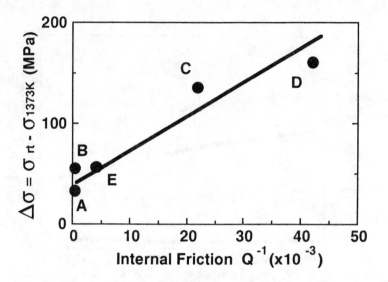

Figure 3. The difference in the flexural fracture strength at room temperature and at 1100°C in alumina ceramics. Specimen designations are given in the figure.

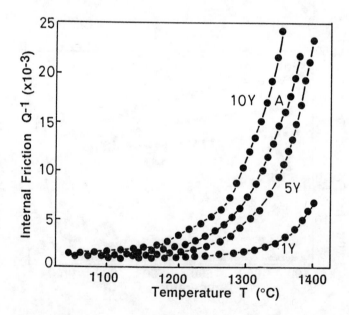

Figure 4. Temperature dependence of the internal friction for alumina ceramics with added yttria at elevated temperatures.
Specimen designations are given in the figure.

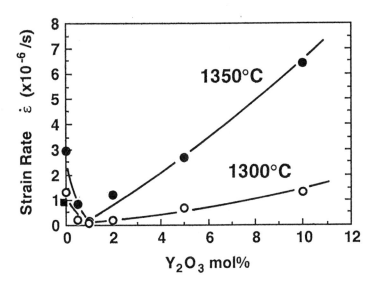

Figure 5.   The dependence of the internal friction at 1300°C and 1350°C on the yttria content.

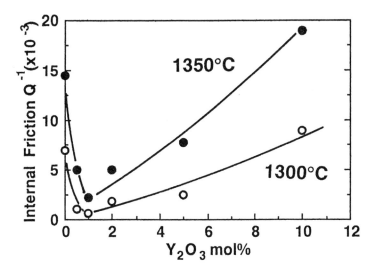

Figure 6.   The dependence of the steady state creep rate at 1300°C and 1350°C on the yttria content.

The stress exponent (n) was measured from the relation between the steady state creep strain rate (ε) and the applied stress (σ), which was varied from 30 to 150 MPa at 1350°C. The n values of specimens A and 1Y were 1.56 + 0.08 and 1.71 + 0.03, respectively. The n value was nearly constant despite the varying additive content. The dependence of the steady state creep rate (ε) on the yttria content is shown in Fig.6. With increasing yttria concentration, ε decrease rapidly up to 1 mol% and then increased gradually. The steady state creep rate of 1Y at 1350°C was 0.8×10^{-6}/s. This value is 25% and 11% of that of specimen A and 10Y, respectively. The curves for the steady state creep rate as a function of yttria content are very similar to the relation of internal friction and yttria concentration.

The microstructures of the specimen 5Y and 10Y are shown in Fig.7. Crystals of YAG were found at the grain boundary when the yttria content was low. With increasing yttria content, the YAG crystals grew reaching the same size as the alumina grains.

DISCUSSION

Degradation of the flexural strength

$MgAl_2O_4$, $Al_6Si_2O_{13}$, and $CaAl_{12}O_{19}$ compounds were formed as a result of the reaction between alumina and MgO, SiO_2, and CaO additives in the specimens, respectively. When magnesia and silica are dissolved in $Al_6Si_2O_{13}$ and $CaAl_{12}O_{19}$, respectively, oxygen ion and cation vacancies can be formed as follows:

$$MgO \xrightarrow{Al_6Si_2O_{13}} Mg'_{Ag} + O_O + 1/2V_O^{\cdot\cdot} \qquad (3)$$

$$SiO_2 \xrightarrow{CaAl_{12}O_{19}} Si_{Ca}^{\cdot\cdot} + 2O_O + 2/3V_{Al}''' \qquad (4)$$

or

$$SiO_2 \xrightarrow{CaAl_{12}O_{19}} Si_{Al}^{\cdot} + 2O_O + 1/2V_{Ca}'' \qquad (5)$$

If titanium oxide is added as a solute in $Al_6Si_2O_{13}$ containing MgO, the concentration of oxygen ion vacancies could be reduced as a result of the conservation of the electronic charge as follows:

$$MgO + Ta_2O_5 \xrightarrow{Al_6Si_2O_{13}} Mg_{Si}'' + 2Ta_{Si}^{\cdot} + 6O_O \qquad (6)$$

An internal friction peak was observed only when oxygen ion or cation vacancies could be formed in the compounds similar to specimens C and D.

The internal friction peaks in specimens C and D result from the relaxation of the applied stress due to stress induced diffusion of ion vacancy-solute atom pairs in the ceramics. The internal friction peak observed in zirconia and Sc_2O_3-CeO_2 depends on the relaxation of Vo-dopant pair[6,7]. The activation energy of the internal friction peaks in specimens C and D was calculated by the peak shift method. The values were 330 and 690 kJ/mol, respectively. These activation energies correspond to those associated with the diffusion of oxygen ions in $Al_6Si_2O_{13}$ and the diffusion of calcium ions in $CaAl_{12}O_{19}$. As mentioned above, the ion vacancy concentration is low in specimens A and B because there are no oxides as solutes in $CaAl_{12}O_{19}$ and $MgAl_2O_4$.

The degradation of the fracture strength depends on the ion vacancies formed in the grain boundary constituents, i.e, the degradation of the strength was larger when ion vacancies were formed in the compounds that formed at the grain boundaries, such as $Al_6Si_2O_{13}$ and $CaAl_{12}O_{19}$ in specimens C and D, than when no vacancies were formed in $Al_6Si_2O_{13}$ in specimen E. When ion vacancies are formed in the grain boundary constituents in the ceramics, they can be deformed as a result of the diffusion of ion and cation vacancies. The applied stress is relaxed by the diffusion of the vacancies. Ion vacancies are then concentrated in the compounds, and the concentrated vacancies form micropores at the grain boundaries. Ultimately, the presence of these micropores will cause the ceramics to fracture.

Steady state creep rate and internal friction

The internal friction, as shown in Fig.4, relates to grain boundary softening. The increment of internal friction in the ceramics at elevated temperatures was caused either by the movement of dislocations in the grain or by viscous flow at the grain boundaries. When internal friction depends on the movement of dislocations, stress amplitude dependency was observed in the internal friction. The internal friction in yttria containing alumina ceramics above 1200°C was caused by viscous flow at the grain boundaries because strain amplitude dependency was not observed in the internal friction at 1250°C.

The activation energy for steady state creep deformation was 419 + 18 kJ/mol. This was calculated from the slope of a plot of ε-1/T. It corresponds to the grain boundary diffusion for aluminium ions (420 kJ/mol) reported by Cannon[8]. The stress exponent n=1 was also reported by Cannon and Chokshi[9,10]. They reported that the creep deformation of polycrystalline alumina corresponded to the diffusion creep at the grain boundary. The n of about 1.5 obtained in this work may correspond to the diffusion creep at the grain boundary. The relation between the internal friction and the steady state creep rate are represented in Fig. 8. The steady state creep rate was proportional to the internal friction.

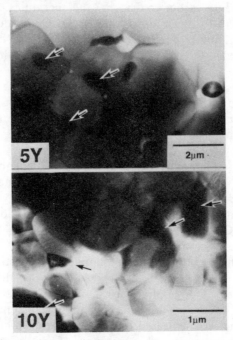

Figure 7.   Microstructures of specimens 5Y and 10Y.
Arrows indicate the YAG crystals.

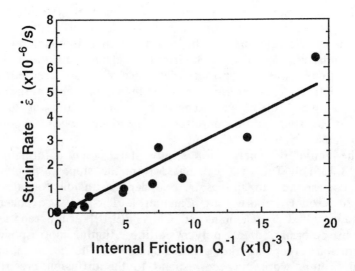

Figure 8.   The relation between internal friction and the steady
state creep rate.

Therefore, the cause of internal friction could be the same as that of creep rate. As mentioned above, the internal friction and creep deformation above 1230°C could be caused by the grain boundary viscous flow due to the diffusion of aluminium ions through the grain boundaries. The internal friction and steady state creep rate become minimum at an yttria content of 1 mol% as a consequence of the fine YAG crystals formed at the grain boundaries, which obstruct grain boundary viscous flow and grain boundary diffusion.

CONCLUSION

(1)   In ceramics, the internal friction at elevated temperatures is related to the degradation of the strength of the grain boundary constituents.

(2)   The presence of ion vacancies was detected easily by an internal friction method.

(3)   The reduction of the ion vacancy concentration is very important for improving mechanical properties at elevated temperatures.

(4)   The steady state creep rate corresponds to the internal friction because the cause of internal friction and creep deformation at elevated temperatures is viscous flow at grain boundaries.

(5)   The dispersion of double oxide particles such as YAG, hindered grain boundary softening, and thus was very effective in improving creep resistance.

REFERENCES

1)   I, Tanaka, G. Pezzotti, K. Matsushita, Y. Miyamoto and T.Okamoto; J.Am.Ceram.Soc.,74(1991),752.

2)   I, Tanaka, G. Pezzotti, K. Matsushita, Y. Miyamoto and T.Okamoto; J.Jap.Appl. Phys., Lattice defects in ceramics (1989),69

3)   M. Shimada, K.Matsushita, S.Kuratani, T.Okamoto, M.Koizumi, K.Tsukuma and T.Tsukidate;J.Am.Ceram.Soc.,67[2](1984),c-23.

4)   K. Matsushita, T. Okamoto and M. Shimada;J.de Phyique,46(1985)C10-549.

5)   K. Matsushita, S. Kuratani, T. Okamoto and M. Shimada;J. Mat. Sci. Lett.,3(1984),345.

6)   R.C.Anderson, F.Z.Noor and A.S.Nowick;Solid State Ionics,9(1983),931.

7)   A.S.Nowick;J.de Phys.,c10(1985),507.

8)   R.M.Cannon, W.H.Rhodes and A.H.Heuer;J. Am. Ceram. Soc., 63(1980),46.

9)   A.H.Chokshi, J.R.Porter; J. Mater. Sci., 21(1986),705.

10)   W.R.Cannon and T.G.Langdon; J. Mater. Sci., 23(1988),1.

THE IMPORTANCE OF PORE-GRAIN BOUNDARY EVOLUTION IN SINTERING KINETICS

G. Q. (Max) Lu and L. X. Liu
School of Mechanical & Production Engineering
Nanyang Technological University
Nanyang Avenue, Singapore 2263

ABSTRACT

This paper presents a theoretical analysis of the pore-grain boundary evolution during sintering. A random pore model is used to study the effects of initial pore structure of a powder compact on the boundary evolution behaviour. The pore size distribution is also studied by assuming a typical Gaudin-Meloy distribution. It is demonstrated that a compact with low initial porosity and high surface area would have fast densification rate. In terms of pore volume distribution, a narrow distribution or high exponent m would result in fast densification. In practical systems, a wide particle size distribution with small mean size of a powder compact is desired to enhance the sintering kinetics and achieve high densification.

INTRODUCTION

Sintering as a primary forming method plays a vital role in ceramics processing. One of the important factors determining the properties of the sintered products is the pore structure as a function of the sintering time. The initial compact matrix and transient pore structure is closely related to the volume densification rate and the grain growth rate during sintering. Due to the pore-grain boundary interactions and diffusion, these two kinetic processes tend to be coupled and mutually inhibiting each other. It is difficult to elucidate such complex kinetics of sintering without a proper model for the pore structure. Most of the models for sintering available in the literature so far used many unrealistic assumptions about the pore structure such as uniform pore with an average size and pore structure based on regular packing of individual grains.

The objective of this study is to develop a more realistic pore structure model for the sintering kinetics and also to review the mechanisms underlying the pore-grain boundary phenomena during sintering. A random pore model is developed based on the fact that most of the ceramic compacts are randomly packed and thus possess a randomly distributed pore structure. The effects of initial pore structure of a powder compact are studied using this random pore model.

THEORY

Mechanism of Sintering Kinetics

The driving force of sintering is the excess free energy of the surfaces of pore-grain boundary. Thermodynamically, the change in free energy of a sintering system can be written as

$$\delta E_{system} = \delta \int \gamma_{pg} \, dA_{pg} + \delta \int \gamma_{gg} \, dA_{gg} \tag{1}$$

where γ_{pg} and γ_{gg} are the energy on the pore-grain surface and grain-grain interface. A_{pg} is the surface area of the pore-grain boundary and A_{gg} is the interfacial area of the grain-grain boundary. In equation (1), the first term on the right hand side is negative since the surface area (A_{pg}) decreases. The second term is positive since the grain boundary area (A_{gg}) increases. As long as $\delta E_{system} \leq 0$, a driving force for sintering exists. One would expect that the free energy of the system will be lowered as the pore-grain boundary is replaced by the grain-grain boundary. This means that at the last stage of the sintering, the driving force for sintering is almost diminished.

In kinetics, the mechanism of sintering is basically dominated by mass transport in terms of various diffusion processes. Several diffusion mechanisms proposed to account for the observed sintering phenomena include vapour transport, volume, surface and grain boundary diffusion. Each of these processes represents a valid means of providing mass transport in a powder compact, and each has been shown to dominate under certain conditions. However, it is generally accepted that, in most practical systems, the kinetics is essentially controlled by diffusion. There are various mathematical expressions for the diffusion kinetics of sintering, which have been reviewed by Kuczynski [1], Coble [2] and others [3, 4].

In general, for several pore-grain boundary configurations, the rate of pore shrinkage (densification) can be described by the following general expression:

$$\frac{dr}{dt} = -\frac{k_p \, D_v \, \gamma_{pg} \, \Omega}{r^2 \, B_o \, T} \tag{2}$$

where D_v is the vacancy diffusion coefficient, (m^2/s); B_o is the Boltzmann's constant and Ω is the atomic volume; T is the temperature and k_p is a constant related to the pore shape.

Although pore densification is the desired and most important process during sintering, grain growth also plays a vital role at some stage of the sintering. The widely accepted theory on the stages of sintering is the one of Coble [2]. According to this theory, the transformation of a powder compact to a solid mass is divided into three stages, namely,

(1)   particle necking stage when necks grow between adjacent particles
(2)   pore shrinkage and grain growth stage
(3)   final densification stage when residual porosity is slowly eliminated.

Stage 1 normally results in little densification and theories for description of this stage is well established. Pore shrinkage during stages 2 and 3 can be described by using equation (3). Whereas, the grain growth rate can be expressed by the grain-grain boundary diffusion theory [6] in terms of the velocity of the grain boundary motion

$$\frac{dG}{dt} - 2V_b - 2F_b \left[\frac{M_p \, M_b}{M_p + nM_b} \right] \tag{3}$$

where M_b is the mobility of the grain-grain boundary and constant at given temperature. and F_b is the force on a grain-grain boundary owing to its curvature and is approximated by

$$F_b - 2w\gamma_{gg}\left(\frac{1}{G} - \frac{1}{G_{max}} \right) \tag{4}$$

where w is a constant and G_{max} is the maximum grain size.

The mobility of a pore-grain boundary M_p is determined by

$$M_p - \frac{D_s \, \delta_s \, \Omega}{B_o \, T \, \pi \, r^4} \tag{5}$$

By defining the boundary mobility ratio,

$$\beta - \frac{n \, M_b}{M_p} \tag{6}$$

We have,

$$\frac{dG}{dt} - \frac{2 \, w \, \gamma_{gg} \, M_b}{(1+\beta)}\left[\frac{1}{G} - \frac{1}{G_{max}} \right] \tag{7}$$

or

$$\frac{dg}{dt} - \frac{2w\gamma_{gg}M_b}{(1+\beta)}\left[\frac{1}{g} - \frac{1}{g_m} \right] \tag{7a}$$

where $g=G/G_0$ and $g_m = G_{max}/G_0$. Combining equation, (2), (5) and (6) gives

$$\beta - \left[\frac{n \, M_b \, B_o \, T \, \pi}{D_s \, \delta_s \, \Omega} \right] (R - (3Kt)^{1/3})^4 \tag{8}$$

where K is the vacancy diffusion rate and

$$K = \frac{k_p \, D_v \, \gamma_{pg} \, \Omega}{B_o \, T} \tag{9}$$

From equation (8), one can see that when $\beta \ll 1$, grain-grain boundary mobility controls the grain growth. Whilst when $\beta \gg 1$, the movement of pore-grain boundary controls the grain growth rate. This is also demonstrated by Brook [6] in his analysis of pore-grain boundary interactions. With equation (8), one would be able to study the pore-grain/grain-grain boundary evolutions and their effects on the grain growth.

A random pore model for a powder compact
The major difficulty in applying the rate expression sintering kinetics such as equations (2) and (7) to real systems lies in relating the pore structural variables to quantities which are actually measured and controlled such as surface area, porosity and pore size distributions. Unfortunately, none of these appears explicitly in the theoretical expressions. The previous approach has been to relate these quantities to average pore size on the basis of an ideal pore model assuming regular packing of individual particles. Due to the strong dependence of the pore densification rate on the pore size, however, the use of this kind of model is not sufficient to elucidate the kinetics of pore-grain evolution. The following section is to develop a more realistic random pore model based on random packing of particles.

Assume that a powder compact is composed of an assembly of spherical particles having a size distribution. The interparticle space constitutes the porosity of the compact. Because the compact is formed by random packing of the particles its pore space can be visualized to be an assembly of randomly distributed cylinders of a length density $l(r)$. The cylindrical pores have overlaps due to its random distribution in space. Then $l(r)dr$ is the length of pore axes per unit volume in $[r, r+dr]$. The equations for the total length, volume and surface area without accounting the overlapping are respectively

$$L_x = \int_0^\infty l(r) \, dr \tag{10}$$

$$e_x = \pi \int_0^\infty r^2 \, l(r) \, dr \tag{11}$$

$$S_x = 2\pi \int_0^\infty r \, l(r) \, dr \tag{12}$$

According to the random distribution theory of Avrami [7], we have

$$(1 - e) = \exp(- e_x) \tag{13}$$

where ε is the pore space fraction and ε_x is the extended pore space per unit volume of compact (without accounting the overlapping). The relationship between the surface area and porosity is given in [8,9]

$$S = S_x(1 - \varepsilon) \tag{14}$$

where S is the total surface area per unit volume.

Therefore, we can write at the onset of sintering

$$\varepsilon_0 = 1 - \exp\left[-\pi\int_0^\infty R^2 \, l(R) \, dR\right] \tag{15}$$

$$S_0 = 2\pi\int_0^\infty R \, l(R) \, dR.(1 - \varepsilon_0) \tag{16}$$

and at any time

$$\varepsilon_t = 1 - \exp\left[-\pi\int_0^\infty r^2 \, l(r) \, dr\right] \tag{17}$$

$$S_t = 2\pi\int_0^\infty r \, l(r) \, dr.(1 - \varepsilon_t) \tag{18}$$

where

$$r(R,t) = R - R^*(t) \tag{19}$$

Using equation (2), we have

$$R^* = (3Kt)^{\frac{1}{3}} \tag{20}$$

The conservation of total pore length leads to

$$l(r)dr = l(R)dR \tag{21}$$

which gives

$$\varepsilon_t = 1 - \exp\left(-\pi\int_{R^*}^\infty (R^2 - 2RR^* + R^{*2})l(R) \, dR\right)$$
$$= 1 - \exp\left(-\pi\int_{R^*}^\infty \left[R^2 - 2R(3Kt)^{\frac{1}{3}} + (3Kt)^{\frac{2}{3}}\right]l(R) \, dR\right) \tag{22}$$

Now, we define the normalized densification rate as

$$\eta = \frac{\epsilon_0 - \epsilon_t}{\epsilon_0} \tag{23}$$

and it can be expressed in terms of the moments of the length density as

$$\eta = \frac{\exp\left[2L_1(3Kt)^{\frac{1}{3}} - L_0(3Kt)^{\frac{2}{3}} - L_2\right] - \exp(-L_2)}{1-\exp(-L_2)} \tag{24}$$

where L_i is the ith moment of the length function $l(R)$. Similarly, we have the surface area ratio

$$\frac{S_t}{S_0} = \left[1 + \frac{\eta\epsilon_0}{1 - \epsilon_0}\right]\left(1 - \left(\frac{L_0}{L_1}\right)(3Kt)^{\frac{1}{3}}\right) \tag{25}$$

For a nonodispersed distribution, one has

$$\eta = \frac{\exp\left[2\pi R_0\, l_0(3Kt)^{\frac{1}{3}} - \pi\, l_0(3Kt)^{\frac{2}{3}} - \pi R_0^2\, l_0\right]-\exp(-\pi R_0^2 l_0)}{1-\exp(-\pi R_0^2 l_0)} \tag{26}$$

$$\frac{S_t}{S_0} = \left[1 + \frac{\eta\epsilon_0}{(1 - \epsilon_0)}\right]\left(1 - \frac{(3Kt)^{\frac{1}{3}}}{R_0}\right) \tag{27}$$

where R_0 and l_0 can be related to the initial porosity ϵ_0 and surface area S_0 by using equations (15) and (16) as follows

$$R_0 = -2(1 - \epsilon_0)ln(1 - \epsilon_0)/S_0 \tag{28}$$

$$l_0 = S_0/(2\pi(1 - \epsilon_0)R_0) \tag{29}$$

To illustrate the results of the above analysis, some simulations will be also performed using a Gaudin-Meloy distribution function for the initial length density,

$$l(R, 0) = \left(\frac{m\, l_0}{R_m}\right)\left[1-(R/R_m)\right]^{m-1} \tag{30}$$

where R_m is the maximum pore radius and m is a distribution exponent.

Therefore, the initial cumulative pore volume distribution is given by

$$\xi(R,0) = \frac{e(R,0)}{e_0} = \frac{1-\exp\left(-\pi\int_0^R R^2 l(R)dR\right)}{1-\exp\left(-\pi\int_0^{R_m} R^2 l(R)dR\right)} \qquad (31)$$

where $\xi(R,0)$ is the fraction of the actual pore volume of pores smaller than R at the onset of sintering.

Similarly, one can write the cumulative pore volume distribution at any time t as,

$$\xi(r,t) = \frac{e(r,t)}{e_t} = \frac{1-\exp\left(-\pi\int_0^r r^2 l(r)dr\right)}{1-\exp\left(-\pi\int_0^{R_m} r^2 l(r)dr\right)} \qquad (32)$$

Using the relationship (21) and the transformation of integration boundary gives

$$\xi(r,t) = \frac{1-\exp\left(-\pi\int_{R^*}^R (R-R^*)^2 l(R)dR\right)}{1-\exp\left(-\pi\int_{R^*}^{R_m} (R-R^*)^2 l(R)dR\right)} \qquad (33)$$

RESULTS AND DISCUSSION

Effects of initial porosity of a powder compact
The initial porosity is a directly measurable property of a powder compact and is largely dependent on the powder size, size distribution and packing method. A small initial porosity generally characterizes a wider particle size distribution. From the theoretical analysis of the random pore model presented in the last section, it is clear that the initial porosity together with surface area determines the initial pore structure such as the mean pore radius and pore size distribution parameters, which in turn determine the initial pore-grain boundary mobility and the driving force for densification.

The effects of initial porosity on the densification rate, mobility ratio and the surface area changes are studied by simulating a monodispersed pore distribution (see equation 26).

Figure 1 shows the effect of initial porosity on the normalized densification (η) versus time for the sintering of undoped Al_2O_3 at 1600 C. The detailed kinetic parameters used for the simulation is listed in Table 1. [10]

It is seen from Figure 1 that the densification rate is very much dependent on the initial porosity. The smaller of the initial porosity the faster the densification. This is because the

TABLE 1. Parameters used for simulations of sintering kinetics of undoped Al$_2$O$_3$

D_s=0.5x10^{-10} m^2/s	D_v=9.8x10^{-16} m^2/s
B_0=1.38x10^{-23} J /K	δ_s=3x10^{-10} m
w=5/4	g_{max}=2.0
Ω=2.11x10^{-29} m^3	M_{b}=2x10^{-14} m^3 /N/s
Γ_{gg}=0.45 J/m^2	Γ_{pg}=0.9 J/m^2

unit pore volume surface area is larger and hence the pore-grain boundary surface free energy is larger provided the initial surface area is constant. When the initial porosity is 0.2 the time needed to reach 80% of densification is only about 1/5th of that in the case of ε_0 =0.4.

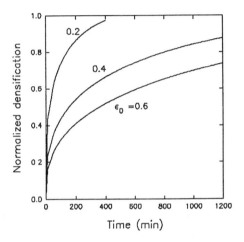

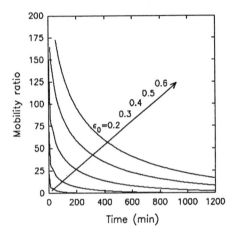

Figure 1. Effect of initial porosity on the densification rate (S_0=1.0x10^6 m^2/m^3)

Figure 2. Effect of initial porosity on the mobility ratio (S_0=1.0x10^6 m^2/m^3)

The boundary mobility ratio ß is the ratio of the grain-grain (g-g) boundary mobility to the pore-grain (p-g) mobility. As discussed earlier, the parameter indicates the control regime for the grain growth. When ß >>1, the movement of pore-grain boundary dominates the grain growth. In this case, the overall densification is largely due to pore-grain boundary movement. As the sintering proceeds, the mobility ratio is decreased to near 1 or below 1 as a result of the replacement of p-g boundary by g-g boundary. The g-g boundary mobility starts to dominate now. This evolution is demonstrated in Figure 2 where the effect of initial porosity is also shown. It shows that for a compact with large porosity, the transition from pore control regime to grain control regime is rather slow. Whereas at low porosity, the compact has relatively low initial mobility ratio and the pore space is drastically reduced once the sintering starts. From Figure 2, one can see that for the case of ε_0 =0.6, the mobility ratio ß will never get below 1 in a practical system. This means that for a compact with large initial porosity it is difficult to achieve total pore elimination by sintering.

The evolution of boundary mobility as discussed above can also be explained by the fact that low porosity compact has large surface area per unit pore volume and thus has larger driving force.   This is further demonstrated in Figure 3.   This Figure shows the evolution of surface area as a function of initial porosity.   As expected the surface area decreases rapidly for low porosity compact.

Effect of initial surface area
As seen from equation 1, the driving force of sintering is dependent on the pore-grain boundary area.   If the system has a large p-g boundary surface area, it will have a large surface energy change during sintering.   Therefore, there would be a fast densification kinetics.   This effect of initial surface area is clearly shown in Figure 4 where the normalized densification rate is plotted against the sintering time.

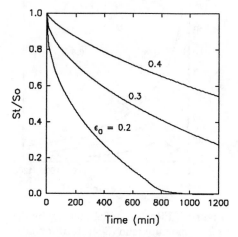
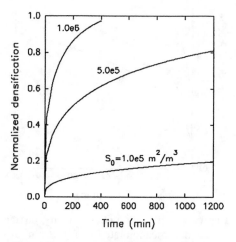

Figure 3. Surface area evolution -effect of initial porosity (S_0=1.0x10^6 m^2/m^3)

Figure 4. Effect of initial surface area on the densification rate (ε_0=0.2)

It is seen that there is a great change in densification rate when the initial surface area is doubled.   In the green compact, a small mean size of the particles would mean a high surface area.   Therefore, a powder having small mean size (high S_0) and wide size distribution (low ε_0) would be desirable for fast sintering and high densification rate.   In figure 5, we have the boundary mobility ratio evolution curves as affected by initial surface area.   It shows that the effect of S_0 behaves similarly to that of initial porosity.   A high initial surface area is expected to facilitate the transition from pore control grain control.   It is seen from Figure 5 that the effect of initial surface area is more pronounced than that of initial porosity.

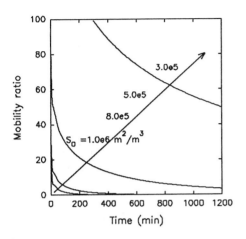

Figure 5. Effect of initial surface area on the boundary mobility ratio(ε_0=0.2)

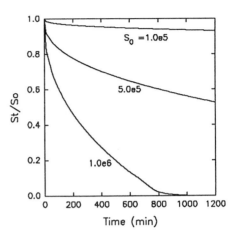

Figure 6. Surface area evolution -Effect of initial surface area (ε_0 =0.2)

Figure 6 show the normalized surface area evolution. It is clear that high initial surface area would also enhance the p-g boundary surface reduction, which is quite understandable from the surface energy point of view.

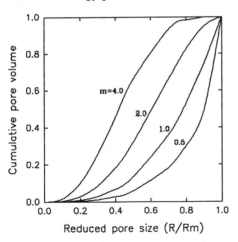

Figure 7. Initial pore volume distribution (Gaudin-Meloy function of pore length)

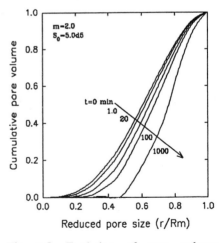

Figure 8. Evolution of pore volume distribution at (m=2.0, S_0=5x10^5 m²/m³)

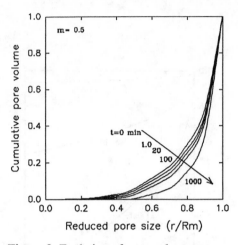

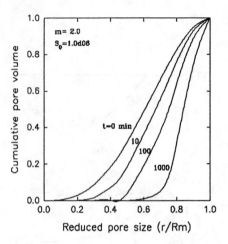

Figure 9. Evolution of pore volume distribution (m=0.5)

Figure 10. Evolution of pore volume distribution (m=2.0, S_0=1x10^6 m^2/m^3)

Effect of pore length density distribution

We assume the density function of pore length follow a Gaudin-Meloy distribution (Eq. 30). The pore volume distribution is then calculated using the random pore model presented in the theory section. Figure 7 shows the initial pore volume distributions for different values of the distribution exponent m. It is seen that as m increases the distribution becomes more convex from a concave shape. The physical meaning of m is that when m >> 1, the small pores constitute the majority of the pore volume and, when m << 1, there are mainly large pores.

The effects of densification during sintering on the pore volume distributions are illustrated in Figures 8 and 9 for m=2.0 and 0.5, respectively. The principal effect of sintering is that, as densification occurs, the pore volume distribution becomes narrower and the median size shifts slowly to larger sizes as more and more fine pores are eliminated. This effect is also observed by Hogg and Hwang [5] for logarithmic pore size distribution.

As there are mainly large pores left in the last stage of the sintering, it is expected to be difficult to eliminate them. With m=0.5, it is noted from Figure 9 that the densification rate is slower than that for m=2.0 simply because a large portion of the initial pores are large pores for m=0.5. A small m behaves like high initial porosity in this case. Comparing Figures 8 and 10, one would find that the effect of initial surface area on the pore volume distribution evolution. Figure 10 is the pore volume evolution curves for S_0=1x10^6 m^2/m^3 being twice of the S_0 value as in Figure 8. It is shown that at high initial surface area, the distribution curves shift faster to the right as sintering proceeds. This further confirms the effect of surface area on the densification rate as discussed earlier (Figure 4).

As to the effect of m on the densification rate, Figure 11 shows the densification curves for various m values. A high m value would result in high densification rate because of high portion of small pores.

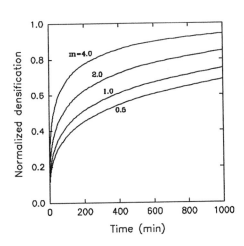

Figure 11. Effect of distribution exponent m on the densification rate

SUMMARY

In this study, a random pore structural model is developed for the powder compact sintering. Based on the popular surface and volume diffusion kinetic theory of sintering, the boundary mobility ratio and the transition of control regime are analyzed using this model. It is shown that the sintering can classified into two stages in terms of boundary evolution, namely, the pore control stage and grain control stage. In the former, sintering is dominated by pore-grain boundary movement, whilst in the later grain growth plays a more important role. Simulations have been performed to study the effect of the initial pore structural properties such as porosity and surface area. It has been demonstrated that a small mean pore size and wide size distribution corresponding to high surface area and low porosity are desirable for fast densification or control regime transition. The simulation results on the effect of pore length density distribution have shown that a convex shape (high m) of the pore volume distribution is more favourable in achieving fast densification.

REFERENCES

1   Kuczynski,G.C., Physics and Chemistry of sintering, <u>Advan. Colloid Interf. Sci.</u>, 1972, **3**, 275

2.   Coble, R.L., Sintering crystalline solids I: Intermediate and final stage diffusion models, <u>J Appl. Phys.</u>, 1961, **32**, 787-792

3.   Yan, M. F. and Cannon, Jr., R. M. and Bowen, H.K., Effect of grain size distribution on sintered density, <u>Mater. Sci. Eng.</u> 1983, **60**, 275-281

4.   Zhao, J. and Harmer, M.P., Effect of pore distribution on microstructure development: I, Matrix pores, <u>J Am. Ceram. Soc.</u>, 1988, **71**, 113-120

5.   Hogg, R. and Hwang, C. L., Pore size distributions in the sintering and heat treatment of agglomerated particulate, in Agglomeration 77, ed. Sastry, K.V.S., AIMMPE, New York, 1977, 198-215.

6.   Brook, R.J., Pore-grain boundary interactions and grain growth, J. Am. Ceram. Soc., 1969, 52, 56-57

7.   Avrami, M., Kinetics of phase change II. Transformation-time relations for random distribution of nuclei, J. Chem. Phys., 1940, 8, 212-224

8.   Sotirchos, S.V., On a class of random pore and grain models for gas-solid reactions, Che. Eng. Sci., 1987, 42, 1262-1265

9.   Bhatia, S.K. and Permutter, D.D., A random pore model for fluid-solid reactions I: Isothermal, kinetic control, AIChE J, 1980, 26, 379-386

10.   Zhao, J. and Harmer, M.P., Effect of pore distribution on microstructure development: II, First and second-generation pores, J Am. Ceram. Soc., 1988, 71, 530-539

Session III

Structural Ceramics I

THE CORRELATION BETWEEN INTERFACE STRUCTURE AND MECHANICAL PROPERTIES FOR SILICON NITRIDE BASED NANOCOMPOSITES

KOICHI NIIHARA, TAKESHI HIRANO, ATSUSHI NAKAHIRA AND KANSEI IZAKI*
The Institute of Scientific and Industrial Research, Osaka University
8-1 Mihogaoka, Ibaraki-shi, Osaka 567, Japan
*The Central Research Center, Mitsubishi Gas Chemical Co. Inc.
Tsukuba, Ibaraki 300-42, Japan

ABSTRACT

Si_3N_4/SiC nanocomposites were prepared by hot-pressing fine, amorphous Si-C-N precursor powders. Transmission electron microscopic observation revealed that the nano-sized SiC particles are dispersed within the matrix grains and/or at the grain boundaries, depending on the volume fraction of the SiC, the size of the SiC particles, and the sintering conditions. The intragranular nano-sized SiC particles affect the growth of elongated Si_3N_4 grains thereby improving fracture toughness and strength. The intergranular nano-sized SiC particles act to control the grain boundary structure by connecting directly with the matrix Si_3N_4 grains without the interface layer, thereby improving the high-temperature fracture behavior.

INTRODUCTION

Silicon nitride (Si_3N_4) ceramics are promising materials for engineering applications because of their high strength, toughness, and thermal shock fracture resistance [1,2]. They have already been used, for example, as automobile engine components. A wide variety of applications, however, require higher toughness and strength at higher temperatures. Reinforcing Si_3N_4 with silicon carbide (SiC) particles or whiskers is one of the most promising techniques for developing such materials [3-7].

The Si_3N_4/SiC whisker composite system produces a significant increase in fracture toughness, but a decrease in fracture strength [3,4]. Investigation by Lange [5] of the mechanical properties of Si_3N_4 ceramics containing dispersed SiC particles (average size: 5, 9, 32 μm) showed that the fracture energy increased for the largest–particle size dispersion series but room–temperature fracture strength was inferior.   Greskovich et al. [6] reported that the fracture toughness of a Si₃N₄/submicron–SiC composite was independent of the volume fraction of SiC.   On the other hand, the SiC particle and whisker dispersions were found to improve the fracture strength at high temperatures [3–5]. No work has been done that has suc– ceeded in improving both fracture toughness and strength at high tempera– tures.

Niihara and colleagues[7–15] have investigated Si₃N₄/SiC nanocomposi– ite, in which the nano–sized SiC particles are dispersed mainly within the matrix grains.   The purpose of this paper is to describe fabrication processes for Si₃N₄/SiC nanocomposites, their micro and nanostructures, and their mechanical properties.   Special emphasis is placed on under– standing the effects of the interface between the nano–sized SiC disper– sions and Si₃N₄ matrix grains on mechanical properties.

EXPERIMENTAL PROCEDURES

The starting amorphous Si–C–N precursor powders, which are transformed to crystalline fine Si₃N₄/SiC mixed powders during sintering, were prepared by vapor phase reaction at 1000°C of a [Si(CH₃)₃]₂NH + NH₃ + N₂ gas system followed by heat–treatment in a N₂ atmosphere at 1350°C for 4 hr to stabi– lize the powders [7].   The C/N ratio in the amorphous powder was adjusted by controlling the flow rate of the gaseous NH₃.   The resulting powder was almost spherical and homogeneous with an average particle size of 0.2 μm.

The amorphous Si–C–N powders with various C/N ratios were mixed with 8 wt% Y₂O₃ as a sintering aid in a plastic bottle using Si₃N₄ balls plus ethanol for 10 hr.   The dried mixtures were hot–pressed at 1800°C for 2 to 4 hr in N₂ under 34 MPa applied pressure.   As a comparison, 6 wt% Al₂O₃ + 2 wt% Y₂O₃ were also used as sintering aids for some of the sintering tests.

The hot–pressed samples were characterized by X–ray diffraction, optical microscopy, scanning electron microscopy (SEM), transmission elec– tron microscopy (TEM), and high resolution TEM.   Bulk density was meas– ured by the Archimedes' immersion technique.   The fracture toughness was estimated by both indentation microfracture and controlled surface flaw methods.   The fracture strength was estimated by a three point bending test up to 1500°C in air.   The span length and crossed–head speed were 20 mm and 0.5 mm/min, respectively.   The tensile surfaces of specimens (3 x 4 x 40 mm) were perpendicular to the hot–pressing axis and polished to 1 μm with diamond paste.

RESULTS AND DISCUSSION

Si3N4/SiC composites with various C/N ratios were fabricated by hot-pressing fine, amorphous Si–C–N powders. X–ray diffraction analysis revealed that the Si3N4/SiC composites were composed of β–Si3N4, β–SiC, a trace of α–SiC, and grain boundary phases, but were free from other impurity phases such as free Si and C. The composites were nearly full–dense and in good agreement with the rule of mixtures.

A transmission electron microsgraph(TEM) of a Si3N4/10vol% SiC composite is shown in Figure 1. The nano–sized SiC particles are located within the matrix Si3N4 grains: Some intragranular SiC particles are indicated by arrows in this figure. A similar microstructure was also observed for composites with both lower and higher SiC content. From these observations, the Si3N4/SiC composites prepared in this work were confirmed to be nanocomposites, in which the nano–sized SiC particles are dispersed mainly within the matrix grains. The number of intragranular SiC particles increased with the increase of SiC volume fraction up to 25 vol% of SiC. By controlling the sintering conditions, the nano–sized SiC particles also could be dispersed both within the matrix grain and at the grain boundaries. Above 25 vol% SiC, the number of nano–sized particles dispersed at the grain boundaries of Si3N4 increased with increasing SiC content. The TEM observation also revealed that the SiC particles smaller than 0.2 μm are predominantly located within the Si3N4 grains and the larger SiC particles are dispersed at the grain boundaries.

Figure 2 shows a high resolution TEM picture of a Si3N4/SiC nanocomposites. There appear to be no impurity phases at the Si3N4/SiC interfaces. The Si3N4 and SiC are directly bonded without impurity phases. As expected from the use of 8 wt% Y2O3 as a sintering additive, the impurity phases were observed at the Si3N4–Si3N4 grain boundaries. However, no interface layer was observed between the intergranular SiC particles and the Si3N4 matrix grains, as can be seen from the lattice image in Figure 3. The lattice match at the interface appears to be quite good, and every fifth lattice plane of SiC becomes the extra plane. The arrows in the picture indicate that every fifth SiC plane does not connect with the Si3N4 lattice. This is one way to accommodate the lattice misfit between different kinds of ceramic interfaces. This observation about the interface between Si3N4 and SiC located at the grain boundaries of Si3N4, however, was only true for the nanocomposites prepared using Y2O3 as a sintering aid. It was not true for the nanocomposites prepared using 6 wt% Al2O3 and 2 wt% Y2O3 as sintering aids.

The scanning electron microscopy(SEM) observations of fractured and polished/etched surfaces indicated that the morphology of the Si3N4 grains in the nanocomposites was strongly influenced by the SiC disper-

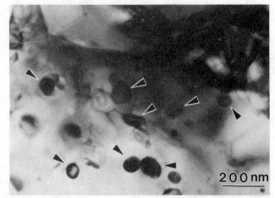

Figure 1   Microstructure of the Si₃N₄/10 vol% SiC nanocomposite (TEM).
Arrows indicate   some of the intragranular SiC particles.

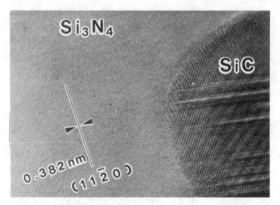

Figure 2   Nano-sized intragranular SiC particle (TEM). The Si₃N₄ is
the (110) plane.

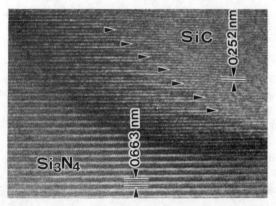

Figure 3   Lattice image of the interface between the intergranular SiC
particles and the Si₃N₄ matrix grains. The arrows indicate
that every 5th plane of the SiC does not connect with the
lattice of the Si₃N₄.

sions. Up to approximately 25 vol% SiC, in which the number of intragranular SiC particles increase with increasing SiC content, the growth of elongated Si3N4 grains was accelerated by the SiC dispersions, as shown in Figure 4. As apparent in figure 4, the elongated Si3N4 grains are observed for both monolithic Si3N4 and the nanocomposite. Compared with monolithic Si3N4, however, the nanocomposites are composed of more uniform and homogeneous elongated Si3N4 grains. At 25 vol% SiC, on the other hand, the growth of elongated Si3N4 grains decreased with increasing SiC content, and fine, equiaxed Si3N4 grains developed. Finally, the nano/nano composites observed above 35 to 40 vol% SiC content could be superplastically deformed at 1600°C [15].

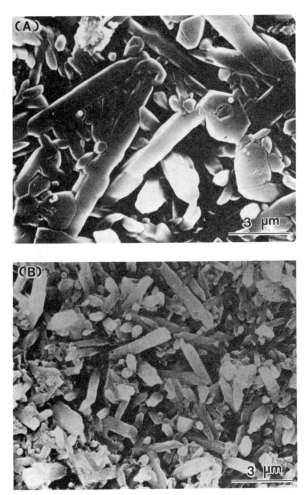

Figure 4 Scanning electron micrographs of the polished surfaces etched by HF solution for Si3N4 (A) and Si3N4/10 vol% SiC nanocomposite (B).

It is reported that elongated Si3N4 grains are formed by solution-diffusion–reprecipitation processes during liquid sintering [15]. From this growth mechanism for elongated Si3N4 grains and TEM observations of the micro and nanostructure change with the nano–sized SiC dispersions, it is reasonable to suggest that the nano–sized SiC particles dispersed within the Si3N4 grains act as nuclei for the growth of β–Si3N4 in the solution–diffusion–reprecipitation process. The decrease of elongated grains observed at higher SiC content can thus be explained by the excess of nuclei sites for β–Si3N4 in these processes. In other words, the intragranular SiC particles are believed to be trapped in the solution–diffusion–reprecipitation process that occurs during the sintering of Si3N4. Mechanical properties should be improved mainly owing to this morphological change of the Si3N4 grains.

The Vickers hardness increased monotonically with increasing SiC content from 15.1 GPa for monolithic Si3N4 to 19.2 GPa for the nanocomposite including 32 vol% SiC, as expected from the rule of mixtures for conventional composites. However, the fracture toughness increased with increasing SiC content up to approximately 25 vol% SiC and then decreased with increasing SiC content, as indicated in Figure 5. The observed maximum value of toughness is 6.7 MPam$^{1/2}$ at 25 vol% SiC, and the lowest toughness value is 5.4 MPam$^{1/2}$ at 32 vol% SiC. But this lowest value observed is still higher than that for monolithic Si3N4. As shown in Figure 5, the fracture strength was also improved by the nano–sized SiC dispersion as expected from linear fracture mechanics. A maximum strength of over 1.5 GPa was achieved at 25 vol% SiC.

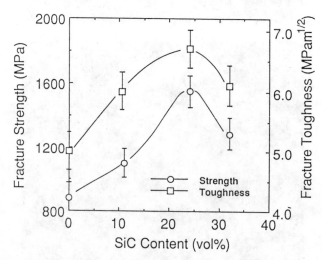

Figure 5 Variation of fracture toughness and strength with the volume fraction of SiC for Si3N4/SiC nanocomposites sintered using 8 wt% Y2O3 as a sintering aid.

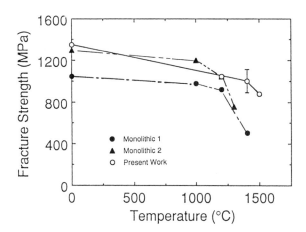

Figure 6　The temperature dependence of 3-point bending strength for the Si3N4/32 vol% SiC nanocomposite fabricated using 8 wt% Y2O3 as a sintering additive. The data for monolithic Si3N4 are also indicated for comparison.

　　Lange [5] fabricated Si3N4 based composites containing 5, 9 and 32μm SiC and found that the fracture toughness slightly increased with dispersions of the larger SiC particles but the strength decreased. Greskovich [6] reports that submicron SiC dispersions did not improve fracture toughness. In this work, on the other hand, both fracture toughness and strength were improved by the dispersion of nano-sized SiC particles. As discussed above, this improvement of toughness and strength is attributed to the promotion by the nano-sized SiC dispersions of the growth of fine and uniformly elongated Si3N4 grains. It is well known that elongated Si3N4 grains improve the fracture toughness of Si3N4 ceramics by the mechanism of crack deflection[16].

　　The fracture strength of Si3N4 fabricated using sintering aids such as Al2O3/Y2O3 and Y2O3 decreases significantly at temperatures above 1200°C because of grain boundary sliding and/or cavitation [8]. Similar behavior was also observed for the monolithic Si3N4 ceramics prepared in this study. As shown in Figure 6, however, the nanocomposite did not exhibit a large degradation in strength up to 1400°C. The fracture strength at 1400°C was over 1000 MPa, and about 900 MPa even at 1500°C for the 32 vol% SiC nanocomposite in which the nano-sized SiC particles are dispersed not only within the Si3N4 grains but also at the grain boundaries.

　　This remarkable improvement of high-temperature strength can be explained by the observation that the nano-sized SiC particles dispersed at the grain boundaries bond directly with the Si3N4 grains (Figure 3). The strong bonding between Si3N4 and SiC decreases the slow crack growth resulting from softening of the grain boundary phases of Si3N4, thereby retaining high strength at high temperatures. An explanation for why nano–sized SiC dispersions have clean interfaces in the presence of rela- tively large amounts of liquid phases is now under investigation.

CONCLUSIONS

Si_3N_4/SiC nanocomposites, in which fine SiC particles are dispersed within the matrix grains, were successfully prepared by hot–pressing Si–C–N precursor powders made using CVD. From the nanostructure–properties relationship, the following important results were obtained.

1) Below 25 vol% SiC, the nano–sized SiC particles are mainly dis- persed within the Si_3N_4 matrix grains where they promote the growth of elongated Si_3N_4 grains. The fracture toughness and strength increase with increasing SiC content.

2) Above 25 vol% SiC, the elongated Si_3N_4 grains decrease and equiaxed grains increase. Toughness and strength then decrease with in- creasing SiC content.

3) Finally, nano/nano composites were obtained with greater than 35 to 40 vol% SiC content that can be superplastically deformed at 1600°C.

4) The morphology changes in the Si_3N_4 grains are attributed to nano–sized SiC particles acting as the nuclei of elongated grains in the solution–diffusion–reprecipitation processes that occur during the sintering of Si_3N_4.

5) The SiC particles dispersed at the grain boundaries of Si_3N_4 bond directly with Si_3N_4 grains without impurity phases. High–temperature strength is strongly improved because the slow crack growth decreases owing to this strong bonding between SiC particles and Si_3N_4 grains.

6) Nano–sized SiC dispersions act both to control the morphology of Si_3N_4 and to control the grain boundary structure.

ACKNOWLEDGEMENT

A part of this was supported by the Japan Ministry of Education under a Grant–in–Aid for Scientific Research (Nos. 03555157 and 03205090).

REFERENCES

1. Godfrey, D.J., Met. Mater., 1968, 2, 305.
2. Edington, J.W., Powd. Met. Int., 1975, 7, 82.
3. Buljun, S.T., Baldoni, J.G. and Huckabee, M.L., Ceram. Bull. 1987, 66, 347.
4. Shalek, P.D., Petrovic, J.J., Hurley, G.F. and Gac, F.G., Ceram.

Bull. 1986, 65, 351.

5. Lange, F.F., J. Amer. Ceram. Soc., 1973, 56, 445.

6. Greskovich and Palm, J.A., J. Amer. Ceram. Soc., 1980, 63, 597.

7. Izaki, K., Hakkei, K., Ando, K., Kawakami, T. and Niihara, K., Ultrastructure Processing of Advanced Ceramics, edited by J.M. Mackenzie and D.R. Ulrich, John Wiley & Sons, Inc., New York, pp.891–900, 1988.

8. Niihara, K., Hirano, T., Nakahira, A. and Izaki, K., Proc. of MRS Meeting on Advanced Materials, Tokyo, pp.107–112, 1988.

9. Niihara, K., Hirano, T., Nakahira, A., Suganuma, K., Izaki, K. and Kawakami, T., J of Japan Society of Powder and Powder Metall., 1989, 36, 243.

10. Niihara, K., Izaki, K. and Kawakami, T., J. Mater. Sci. Lett., 1990, 9, 598.

11. Niihara, K., Suganuma, K. and Izaki, K., J. Mater. Sci. Lett., 1990, 9, 112.

12. Niihara, K., Izaki, K. and Nakahira, A., J of Japan Society of Powder and Powder Metall., 1990, 37, 352.

13. Niihara, K., New Ceramics, 1989, 2, 1.

14. Izaki, K., Nakahira, A. and Niihara, K., J. of Japan Society of Powder and Powder Metall., 1991, 38, 357.

15. Niihara, K., J. Ceram. Soc. Japan, 1991, 99, 974

16. Tani, E., Umebayashi, S., Kishi, K., Kobayashi, K. and Nishijima, M., Ceram. Bull., 1986, 65, 1311.

α'/β' SIALON COMPOSITES

Yoshio Ukyo and Shigetaka Wada
Toyota Central Research and Development Labs.,Inc.
Nagakute-cho, Aichi-gun, Aichi-ken 480-11, Japan

ABSTRACT

Si_3N_4 ceramics containing Y-α'-sialon were produced by hot-pressing mixtures of Si_3N_4, Y_2O_3, and AlN powders at 1750 to 1850°C under a pressure of 20 MPa for 1 to 10 hours. X-ray diffraction analysis indicated that the hot-pressed bodies were composed of Y-α'-sialon and β'-sialon. The amount α'-sialon increased as the amount of Y_2O_3 and AlN were increased. The lattice constant (solubility) of α'-sialon depended on the amount of Y_2O_3 and AlN, while that of β'-sialon did not. The amount of α'-sialon decreased with increasing sintering temperature. This phenomenon was observed when Si_3N_4 ceramics containing α'-sialon were heated at higher temperatures for a long time. From these results, it is concluded that when α'-sialon coexists with β'-sialon, it is a metastable at high temperatures.

INTRODUCTION

Si_3N_4 exists as two polytypes, α and β. The solid solution of β'-Si_3N_4, in which Al and O substitute for Si and N, respectively, is called β'-sialon. It is represented by the general formula, $Si_{6-z}Al_zO_zN_{8-z}$ ($0<Z\leq 4.2$) [1]. In other words, the Si-N bond is partly replaced by the Al-N bond. It is also known that the Si in β-Si_3N_4 can be substituted by Ga [2] and Be [3]. On the other hand, α-Si_3N_4 forms another solid solution called α'-sialon. In this solid solution, Si and N can be substituted by Al and O, respectively. Moreover, some elemental metals such as Li, Y, Ca, and Mg dissolve interstitially thereby maintaining electroneutrality. α'-sialon is represented by the general formula, $Mx(Si,Al)_{12}(O,N)_{16}$, where M represents Li, Y, Ca, or another of the metal elements.

In an Y-Si-Al-O-N system, α'-sialon containing Y (Y-α'-sialon) forms in the limited range determined by Park et al. [4] as shown in Figure 1. Mitomo et al. [5] obtained Y-α'-sialon by sintering mixtures of Si_3N_4, Y_2O_3, and AlN powders. In this case, the Y-α'-sialon is synthesized by the following reaction:

$$(4-3a)\,Si_3N_4 + a\,(Y_2O_3 + 9AlN) = Y_{9a}\,(Si_{12-9a},Al_{9a})\,(O_{3a},N_{16-3a}) \qquad (1)$$

Therefore, Y-α'-sialon is produced when Y_2O_3 and AlN in the mole ratio of 1:9 are added to Si_3N_4.

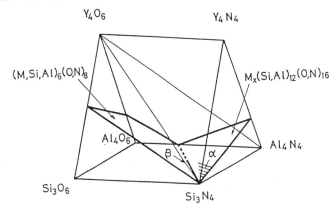

Fig. 1 Outline representation of Y-α'-sialon and β'-sialon

systems.

When Y-α'-sialon is produced by sintering mixtures of Si_3N_4, Y_2O_3, and AlN powders, the system containing Y-α'-sialon is usually represented by the plane containing Si_3N_4, Y_2O_3, and AlN in Figure 2 [5] [6]. The single-phase Y-α'-sialon region is the intersection of the planes containing Si_3N_4–YN·3AlN–4/3(Al_2O_3·AlN) and Si_3N_4–Y_4O_6 (Y_2O_3)–Al_4N_4 (AlN). Therefore, this region corresponds to the dotted line A, shown in Figure 2. Both β-Si_3N_4 and Y-α'-sialon form on the straight line B between Si_3N_4 and the Y-α'-sialon phase.

Characterization of sintered Si_3N_4 consisting of Y-α'-sialon and β-Si_3N_4 was carried out by Mitomo et al. [5] [7] [8]. They reported that either Y-α'-sialon alone or Y-α'-sialon plus β-Si_3N_4 were obtained by sintering mixtures of Si_3N_4, Y_2O_3, and AlN powders. In their reports, the effect of oxygen on the forming reaction of α'-sialon was not taken into account. However, oxygen usually exists in the Si_3N_4 powder as a surface oxide film. The formation of Y-α'-sialon appears to be influenced by the oxygen.

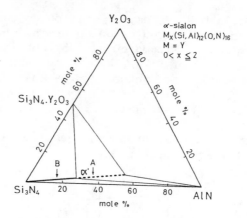

Fig. 2 $Si_3N_4-Y_2O_3-Al_2O_3$ section of Si-Y-Al-O-N system at 1750°C.

The purpose of this study is to investigate the formation of Y-α'-sialon by sintering mixtures of Si_3N_4, Y_2O_3, and AlN powders with various compositions, the effect of oxygen on the formation of Y-α'-sialon, and the stability of Y-α'-sialon at high temperatures.

EXPERIMENTAL PROCEDURE
Raw powders
The results of chemical analysis of the raw powders used in this experiment are shown in Table 1. Ten kinds of mixtures with different total amounts of the additives, Y_2O_3 and AlN, were prepared. In all the mixtures, the mole ratio of Y_2O_3 to AlN was 1 to 9, as shown in Table 2. According to the results obtained by Mitomo et al. [5], the composition of mixtures 1 through 8 correspond to the region in which the Y-α'-sialon coexists with β-Si_3N_4, and the compositions of mixtures 9 and 10 correspond to the single phase region of Y-α'-sialon in Figure 2. These powders were mixed in a polyethylene container with ethanol and Si_3N_4 balls for 20 hours. The mixtures were filtered and then dried at about 80°C for 50 hours. The dried mixtures were die-pressed at 20 MPa. The compacts were about 60 mm in diameter and 10 mm thick.

Hot-pressing
The powder mixtures were hot-pressed in a graphite die coated with BN. Hot-pressing was done in a N_2 atmosphere at 1750, 1800, and 1850°C under a pressure of 20 MPa for 1 to 10 hours.

Evaluation
Specimens measuring 10 x 20 x 2mm were cut from the hot-pressed bodies to determine the phases and to measure their

Table 1 Results of chemical analysis of raw powders

Raw Powders	O	Ca	Fe	C	Al	Si
Si_3N_4	1.3	<50	<100		<50	
AlN	0.9	77	<10	250		<15
Y_2O_3		<0.5	1.8			
	(wt%)			ppm		

Table 2 Nominal compositions of powder mixtures (mol%)

No.	Si_3N_4	Y_2O_3	AlN	x *
1	91.0	0.9	8.1	0.077
2	89.0	1.1	9.9	0.096
3	87.0	1.3	11.7	0.115
4	85.0	1.5	13.5	0.135
5	83.0	1.7	15.3	0.150
6	81.0	1.9	17.1	0.174
7	77.0	2.3	20.7	0.215
8	74.0	2.6	23.4	0.255
9	67.0	3.3	29.7	0.355
10	50.0	5.0	45.0	0.622

* x: $Y_x(Si,Al)_{12}(O,N)_{16}$

lattice constants by X-ray diffraction (XRD) analysis. XRD analysis was done using RU-3L type equipment (Rigaku Denki Co. Ltd.) The scanning speed, the accelerated voltage, and the current were 0.25° (2θ) per minute, 40 KV, and 80 mA respectively. The composition of grains and grain boundaries, particularly tripple point, were analyzed by scanning transmission electron microscopy (STEM).

EXPERIMENTAL RESULS

Determination of Phases
    An example of the the XRD patterns obtained is shown in Figure 3. Two sets of peaks corresponding to β -Si_3N_4 and α -

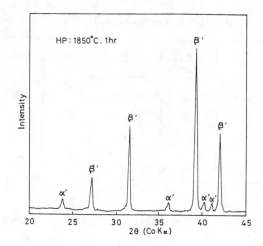

Fig. 3   An example of X-ray diffraction patterns of specimen 2.

Si_3N_4 were found in the XRD patterns of specimen 1 through 8.
In the XRD patterns of specimens 9 and 10, on the other hand,
peaks corresponding only to α -Si_3N_4 were found.   All peaks were
shifted towards smaller angles, and the crystalline phases were,
therefore, detrermined as Y-α '-sialon and β '-sialon.
According to equation (1), all of the Y, Al, O, and N in the
Y_2O_3 and AlN should be dissolved in the Si_3N_4 structure with the
formation of Y-α '-sialon.   Actually, however, β '-sialon was
formed.   Nevertheless, the mole ratio of Y_2O_3 to AlN was 1:9/
This means that some of the Al and O in the Y_2O_3 and AlN were
consumed for the formation of β '-sialon rather than Y-α '-
sialon.

(α ':β ') ratio
      The ratios of the Y-α '-sialon to the β '-sialon were
determined from the peak heights of the XRD patterns using the
following equation:

$$\alpha ' (\%) = [I_{102}(\alpha ') + I_{210}(\alpha ')] \times 100/$$
$$[I_{101}(\beta ') + I_{210}(\beta ') + I_{102}(\alpha ') + I_{210}(\alpha ')] \qquad (2)$$

where $I_{102}(\alpha ')$ and $I_{210}(\alpha ')$ are the peak heights of planes
(102) and (210) of Y-α '-sialons, and $I_{101}(\beta ')$ and $I_{210}(\beta ')$
are the peak heights of planes (101) and (210) of β '-sialon,
respectively.
      Figure 4 shows the dependence of the amounts of Y-α '-
sialon on the amounts of Y_2O_3 and AlN.   The abscissa of this
figure represents the x value in the general formula,
$Yx(Si,Al)_{12}(O,N)_{16}$.   The x values were calculated from the
nominal compositions of the mixtures shown in Table 2.   As

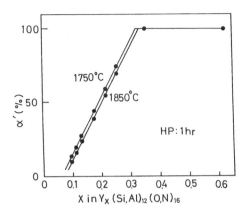

Fig. 4   Dependence of the amount of Y-α'-sialon on x and

hot-pressing temperature.

described above, SiO₂ present in the Si₃N₄ powder as an impurity
seems to affect the formation of α'- and β'- sialon. However,
this effect of SiO₂ was not taken into consideration in the
calculation of x in this experiment.
      The amount of Y-α'-sialon increased with increasing
and AlN. Furthermore, single-phase Y-α'-sialon was formed when
the amount of Y₂O₃ and AlN present greater than 30 mol% (x is
larger than 0.33). The amount of Y-α'-sialon decreased with
increasing hot-pressing temperature. Figure 5 shows the amount
of Y-α'-sialon in the specimens hot-pressed at 1750 and 1850℃
for 1 and 10 hours. The amount of Y-α'-sialon decreased with
increasing hot-pressing temperature and time. These results
lead to the conclusion that the Y-α'-sialon that is formed at
the initial stage of sintering gradually changes into β'-
sialon.

Lattice Constants (Solubility)
      The lattice constants of Y-α'-sialon and β'-sialon were
calculated from the diffraction angles of the XRD peaks. Planes
(210), (200), (110), (031), and (201) of the α-Si₃N₄ (α'-
sialon), and (510), (330), (320), (212), and (411) of the β-
Si₃N₄ (β'-sialon) were used for the determination of the
lattice constants. In the case of Y-α'-sialon, peaks at low
diffraction angles were used because the intensity of peaks at
high diffraction angles is too weak to analyze when the amount
of Y-α'-sialon formed is very small.
      Figures 6(a) and (b) show the compositional dependence of
the lattice constants of Y-α'-sialon and β'-sialon for the
specimens hot-pressed at 1850℃. The lattice constants of Y-

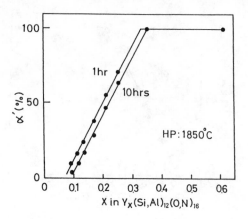

Fig. 5   Dependence of the amount of Y-α'-sialon on the hot-

pressing time.

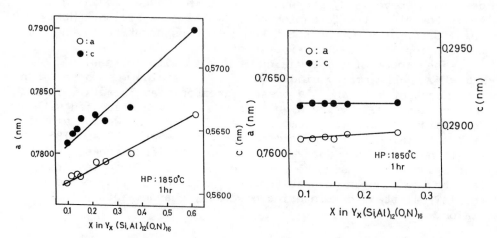

Fig. 6(a)   Relation between lattice constant of Y-α'-sialon
and x.

Fig. 6(b)   Relation between lattice constant of β'-sialon

α'-sialon increased with increasing x (the amount of Y_2O_3 and
AlN) not only when only Y-α'-sialon formed but also when both
Y-α'-sialon and β'-sialon formed.   On the other hand, the
lattice constants of β'-sialon were independent of x.   The β'-
sialon produced is represented by the following formula:

$$Si_{5.4}Al_{0.6}O_{0.6}N_{7.4} \tag{3}$$

Analysis by STEM

Figure 7 shows a microstructure of Si_3N_4 observed by transmission electron microscopy (TEM). The existence of two kinds of grains with differrent morphologies and compositions was revealed from the analysis by STEM/EDS. The grains with different morphologies have different compositions. Table 3 shows the compositions of the grains analyzed. The compositions were calculated on the assumption that the grains were composed of only Si, Y, and Al because O and N could not be analyzed by STEM. These results indicate that grains A and B are β'-sialon and Y-α'-sialon, respectively. The compositions of triple point determined by the same method is also shown in Table 3. Larger amounts of Y and Al were found at the triple point than in the grains. From the observation of the microstructures by TEM, the triple point is assumed to be a glassy phase composed of Si-Y-Al-O-N.

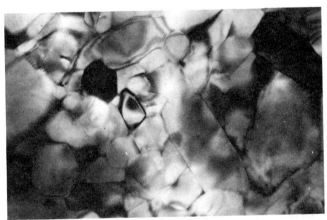

1μm

Fig. 7 Microstructure of specimen hot-pressed at 1850°C

for 1 hr.

Table 3 Compositions of grains and grain boundaries

	Si	Y	Al
A (β'-sialon)	94.62	0.05	5.33
B (α'-sialon)	94.99	0.27	4.74
Triple point	73.09	17.89	9.02

Note: The compositions were calculated on the assumption that the grains were composed of only Si, Y and Al, because O and N could not be analyzed by STEM.

DISCUSSION

When the mole ratio of Y_2O_3 to AlN is 1:9, all of the Y_2O_3 and AlN present can dissolve in the Si_3N_4, and Y-α'-sialon is formed according to equation (1). Both β-Si_3N_4 and Y-α'-sialon are formed when the amount of Y_2O_3 and AlN present is small. On the other hand, only Y-α'-sialon is formed when a large amount of Y_2O_3 and AlN is used.

In this experiment, Si_3N_4 consisting of Y-α'-sialon and β'-sialon was obtained by hot-pressing mixtures of Si_3N_4, Y_2O_3, and AlN powders. Mitomo et al. [5] and Huang et al. [6] reported a different result, namely Y-α'-sialon and β-Si_3N_4 (not β'-sialon) were formed.

As previously discussed, oxgen is presumed to be present in the Si_3N_4 powder as a SiO_2 film on the surface of the Si_3N_4 powder [12]. As shown in Table 1, the oxygen content of the Si_3N_4 powder used was about 1.3 wt% (2.5mol%), that is, (about 5.7mol% as SiO_2). Although only Y_2O_3 and AlN were used as additives, the actual system must be considered to be Si_3N_4-Y_2O_3-AlN-SiO_2 rather than the Si_3N_4-Y_2O_3-AlN, as shown in Figure 2.

Slassor and Thompson [9] determined the behavior diagram that includes both the Y-α'-sialon and β'-sialon phase shown in Figure 8. This diagram correponds to the plane in Figure 1, on which the ratio of metal to nonmetal atoms is maintained at 3:4. The wide region in which Y-α'-sialon and β'-sialon coexist is found in this diagram. The straight line connecting Si_3N_4 with point A between YN·3AlN and 3/4(Al_2O_3:AlN) is identical with the straight line A and B in Figure 2, on which the mole ratio of Y_2O_3 to AlN is 1:9. The actual composition, therefore, deviates from this straight line towards the region where Y-α'-sialon and β'-sialon coexist, on the plane shown in Figure 8.

Kamiya et al. [10] reported that SiO_2 dissolves readily in Si_3N_4 in the presence of AlN. Therefore, some part of the AlN added will be consumed for the formation of β'-sialon as indicated in the following equation:

$$(2-0.5_z)Si_3N_4 + zAlN + 0.5zSiO_2 = Si_{6-z}Al_zO_zN_{8-z} \qquad (4)$$

The result of XRD analysis shows that the lattice constant of Y-α'-sialon in specimen 1 containing a small amount of Y_2O_3 and AlN is slightly larger than that of the α-Si_3N_4 used. This indicates that the solubility of Y-α'-sialon is very poor. However, the solubility of β'-sialon in specimen 1 corresponds to z=0.6. It is postulated that β'-sialon is formed in preference to Y-α'-sialon at the initial stage of sintering, and that the residual AlN which is not consumed for the

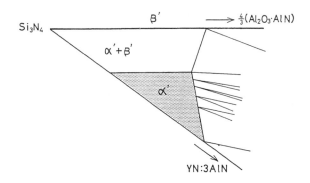

Fig. 8 Simplified Si_3N_4-YN:3AlN-$\frac{4}{3}$(Al_2O_3:AlN) section of

Si-Y-Al-O-N system at 1750°C.

formation of β'-sialon dissolves in Si_3N_4, resulting in the
formation of Y-α'-sialon. Therefore, Y_2O_3 ultimately exceeds
AlN even if the nominal mole ratio of Y_2O_3 to AlN is 1:9. As
discussed in the introduction, Y_2O_3 can only dissolve in the
Si_3N_4 together with AlN. It is postulated that the excess Y_2O_3
reacts with SiO_2 and remaing at the grain boundaries.
    The analysis by STEM showed that a large amount of Al and Y
was present at the grain boundaries and that some part of the Y
and the Al did not dissolve in Si_3N_4, although the mole ratio of
Y_2O_3 to AlN was maintained at 1:9. Bando [11] also reported
that a large amount of Y was present at the grain boundaries in
the Y-α'-sialon. The detailes of the effects of SiO_2 and the
grain size on the formation of Y-α'-sialon and β'-sialon will
be described elsewhere.

    The dependece of the lattice constants (solubility) of Y-
α'-sialon differs from the results obtained by Mitomo et al.
[5]. They reported that Y-α'-sialon coexists with β-Si_3N_4,
that the dependence on x of the lattice constants of Y-α'-
sialon coexisting with β-Si_3N_4 was not observed, and that a
single Y-α'-sialon phase formed when x was larger than about
0.3. It was observed in this experiment, however, that Y-α'-
sialon coexists with β'-sialon, and that the lattice constants
(solubility) of Y-α'-sialon were increased with increasing Y_2O_3
and AlN. A single Y-α'-sialon phase was obtained when x was
larger than about 0.33. This value differs slightly from the
value reported by Mitomo et al. [5]. The reason for this
discrepancy is not yet understood. However, it is postulated
that difference in the characteristics of the starting powder
such as purity and grain size, affect the synthesis of Y-α'-
sialon.

In the diagram determined by Slàssor et al. [9], tie lines were drawn in accordance with the lattice constants (solubility) obtained, as shown in Figure 9. The amounts of Y-α'-sialon and β'-sialon are also shown in the diagram. The compositions of the specimens should fit along the straight line A, when mole ratio of Y_2O_3 to AlN is 1:9. However, the compositions of the specimens obtained in this experiment deviated from line A and appeared to fit along curve B. It is believed that this deviation can be attributed to the presence of SiO_2 in the Si_3N_4 powder.

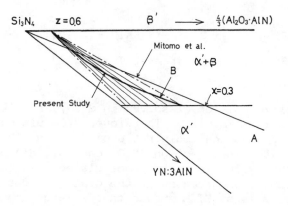

Fig. 9   Compositional changes of specimen in the co-existing

region of Y-α'-sialon and β'-sialon.

The amount of Y-α'-sialon decreased with increasing hot-pressing temperature and time. This suggests that the Y-α'-sialon that coexists with the β'-sialon is unstable and is transformed into β'-sialon with time, at high temperatures. The specimens containing various amounts of Y-α'-sialon fabricated by hot-pressing at 1850℃ for 1 hour were annealed at 1850℃ for 50 hours to investigate the stability of the Y-α'-sialon coexisting with β'-sialon. Figure 10 shows the change in the amount of Y-α'-sialon in the annealed specimen and in the specimen fabricated by hot-pressing at 1850℃ for 10 hours. The amount of Y-α'-sialon decreased with time as shown in Figure 10. It is concluded, therefore, that Y-α'-sialon is less stable than β'-sialon.

CONCLUSIONS

Si_3N_4 composed of Y-α'-sialon only and of Y-α'-sialon plus β'-sialon was fabricated by hot-poressing mixtures of Si_3N_4, Y_2O_3, and AlN powders at 1750, 1800 and 1850℃ for 1 to

123

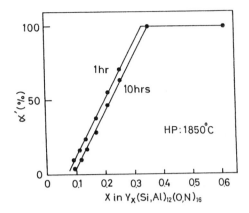

Fig. 10 Dependence of the amount of Y-α'-sialon on the annealing time at 1850°C.

10 hours. The dependece of the amount of Y-α'-sialon, the lattice constants (solubility), and the stability of Y-α'-sialon at high temperatures were studied. The following results were obtained.

1. Both Y-α'-sialon and β'-sialon are formed when the x value in Yx(Si,Al)$_{12}$(O,N)$_{16}$ is smaller than 0.33. When the x value is larger than 0.33, a single Y-α'-sialon phase is formed.

2. The lattice constant (solubility) of Y-α'-sialon depends on the amount of Y$_2$O$_3$ and AlN present, while the lattice constant of β'-sialon dose not change.

3. Y-α'-sialon is tranformed into β'-sialon by heating at high temperatures for a long time. It is concluded that when Y-α'-sialon coexists with β'-sialon, it is metastable at high temperatures.

ACKNOWLEDGMENTS
The authors thak Dr. T. Okamoto for his helpful discussions, and Messrs. T. Ishiguro and N. Suzuki for the STEM analysis.

REFERENCES

1. Jack, K.H., Sialons and Related Nitrogen Ceramics. J. Mater. Sci., 1976, 11, 1135-58.

2. Oyama, Y., Solid Solution in the Ternary System Si$_3$N$_4$-Al$_2$O$_3$-Ga$_2$O$_3$. Japan J. Appl. Phys., 1972, 11, 1572.

3. Huseby, I., Lukasu, H.L. and Petzow, G., Phase Equilibria in the System $Si_3N_4-SiO_2-BeO-Be_2N_3$. J. Amer. Ceram. Soc., 1975, 58, 377-80.

4. Park, H.K., Thompson, D.P. and Jack, K.H., α-Sialon Ceramics. Science of Ceramics, 1980, 10, 251-6.

5. Mitomo, M., Izumi, F., Bando, Y. and Sekikawa, Y., Characterization of α'-Sialon Ceramics. Proc. Int. Symp. on Ceramic Component for Engine, 1983, 377-86.

6. Huang, Z-K., Greil, P. and Petzow, G., Formation of α-Si_3N_4 Solid Solutions in the System $Si_3N_4-AlN-Y_2O_3$. J. Amer. Ceram. Soc., 1983, 66, c96-7.

7. Mitomo, M. and Furukawa, O., The Stability of α'-Sialon at High Temperature. Yogyo-Kyokai-Shi, 1982, 89, 631-3.

8. Izumi, F., Mitomo, M. and Bando, Y., Rietveld Refinements for Calcium and Yttrium Containing α'-Sialons. J. Mater. Sci., 1984, 19, 3115-20.

9. Slasor, S. and Thompson, D.P., Preparation and Characterization of α'-Sialons. Proc. Int. Conf. on Non-Oxide Technical and Engineering Ceramics, 1987, 223-30.

10. Kamiya, N., Oyama, Y. and Kamigaito, O., Silicon Nitride Solid Solution in the Ternary System, $Si_3N_4-AlN-SiO_2$. Yogyo-Kyokai-Shi, 1975, 83, 553-7.

11. Bando, Y., Analysis of Boundary Composition of Sialons by STEM. Ceramics Japan, 1982, 17, 625-7.

THE EFFECTS OF GRAIN SIZE ON STRENGTH, FRACTURE TOUGHNESS, AND STATIC FATIGUE CRACK GROWTH IN ALUMINA.

NOBUYUKI MIYAHARA, YOSHIHARU MUTOH, KOUHEI YAMAISHI,
KEIZO UEMATSU AND MAKOTO INOUE
Nagaoka University of Technology
Kamitomioka,Nagaoka,940-21,Japan

ABSTRACT

Three point bending, fracture toughness and static fatigue crack growth tests of alumina specimens with various grain sizes were carried out to investigate the effects of grain size on bending strength, fracture toughness, and static fatigue crack growth rate. The bending strength increased with decreasing grain size. The validated fracture toughness is independent of grain size. On the other hand, the apparent fracture toughness with slow crack growth increased with increasing grain size due to the R-curve behavior. The static fatigue crack growth rate was reduced with increasing grain size.

INTRODUCTION

It is of importance in material-designing and processing of structural ceramics to understand the relationship between strength, fracture characteristics, and grain size. In metallic materials, the so-called Hall-Petch or Hall-Petch-like relationships [1] between grain size and yield stress, cleavage fracture strength, and fatigue limit can be observed. These relationships occur because the grain boundary obstructs the movement of dislocations and slips. The Hall-Petch-like relationship between strength and grain size in alumina ceramics has been reported by several investigators [2]~[6]. The reason that the Hall-Petch-like relationship is observed in alumina ceramics is not clearly understood. There has only been a small effort made to understand the effects of grain size on such important fracture characteristics in structural ceramic materials as fracture toughness and the property of static fatigue.

In the present study, alumina specimens with various grain sizes were produced using the same fine grain starting powders.

Bending, fracture toughness and static fatigue crack growth tests were carried out to investigate the effect of grain size on these strength and fracture characteristics. Detailed fractographic observation was also carried out to investigate the role of grain boundary on strength and fracture characteristics.

MATERIAL

The starting materials were highly pure (99.9%) and fine grain (0.77 μ m) active alumina powder granules (Taimei Kagaku Kogyo, TM-DS). The granules were molded into bars (80× 15× 15 mm) at 20 MPa and isostatically pressed at 300 MPa. After heating at 600℃ for 2 h to remove the binder, the bars were presintered at temperatures ranged from 1280℃ to 1600℃ for 1 h in air. The presintered bars were consolidated by hot isostatic pressing at temperatures from 1250℃ to 1550℃ at 100 MPa for 1 h using commercial equipment (NKK-ASEA QIH-9). The resulting mean grain diameters of the sintered materials prepared are listed in Table 1 together with the relevant sintering conditions.

TABLE 1
Sintering temperatures and mean grain diameter for alumina specimens tested.

Specimen	Temperature of presintering (℃)	Temperature of HIP sintering (℃)	Mean grain diameter (μ m)
A	1280	1250	0.8
B	1300	1250	1.1
C	1400	1350	2.4
D	1500	1430	3.9
E	1600	1550	5.9

EXPERIMENTAL PROCEDURE

Bending strength test
Bending specimens with dimensions of 35× 4× 3 mm were machined from the sintered bars. The testing surface was polished with successively finer diamond paste to remove the grinding defects and residual stress. The resultant surface roughness Ra was less than 0.1 μ m. Three-point bending tests were carried out using an Instron-type testing machine (Shimadzu DSS-10T) with span length of 30 mm and crosshead speed of 0.5 mm/min.

Fracture toughness test

Controlled surface flaw (CSF) method: A surface crack was introduced on the surface of the bending specimen (35× 4× 3 mm) by using a Vickers indentor. The indented surface was ground up to almost 4 times the depth of the introduced indentation and finally lapped to remove the residual stress induced during the indentation and subsequent grinding [7]. The resultant crack length on the surface 2c was from 450 to 600 μ m and the depth was from 200 to 250 μ m. A three-point bend fracture toughness test was carried out using an Instron-type testing machine with span length of 30 mm and crosshead speed of 0.5 mm/min. The Newman-Raju equation [8] was used for calculating the K-value of a surface crack.

Fatigue precracking (FP) method: A through-the-thickness fatigue precrack was introduced in the fracture toughness specimen (55× 10× 5 mm) according to the following two-step procedure. In the first step, to initiate fatigue-cracking, a pop-in crack with a crack length of from 2 to 3 mm was introduced by the Bridge Compression (BC) method [9], in which starter indentation cracks are induced on the center line of a specimen surface and the lower anvil with a groove width of 7 mm is used as shown in Fig.1. A fatigue crack was extended up to the total precrack length of 4 mm from the tip of the BC precrack using a servohydraulic fatigue machine (Simadzu EHF-F1) under the conditions of three-point bending, with a span length of 40 mm, a stress ratio of 0.1, and a frequency of 20 Hz. The maximum applied load for cyclic fatigue cracking was limited to a maximum stress intensity factor, K_{fmax}, lower than 1.5 MPa$\sqrt{}$ m, which was low enough to satisfy the requirement for fatigue precracking of the ASTM standard for fracture toughness test method, E399. The three-point bend fracture toughness test was carried out using an Instron-type test machine with span length of 40 mm and crosshead speed of 0.5 mm/min.

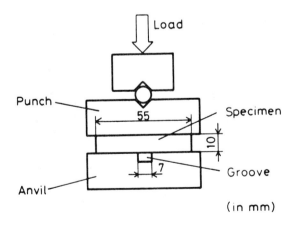

(in mm)

Figure 1. Schematic illustration of the BC method.

Static fatigue crack growth test

For the static fatigue crack growth test, specimens with dimensions of 55× 10× 5 mm were precracked by the BC method [9]. The length of precracks ranged from 2 to 3 mm. A constant load was applied to the specimen in three-point bending with a span length of 40 mm. A travelling microscope with an accuracy of 0.01 mm was used to measure the crack length. A K-decreasing test was also carried out to obtain the threshold value of static fatigue crack growth, K_{Iscc}. The threshold condition was defined when no crack growth was observed for 24 h.

RESULTS

Bending strength

It is known from the experimental results reported that the bending strength of polycrystalline ceramic materials depend on the mean grain diameter. The relationship between grain size and strength is expressed by the following Hall-Petch-like equation [1].

$$\sigma_f = \sigma_\theta + kd^{-1/2} \tag{1}$$

where, σ_θ, and, k, are material constants and independent of grain size.

From the present results, the relationship between bending strength, σ_f, and the inverse of the square root of grain diameter, $d^{-1/2}$, is shown in Fig.2 by using the symbol of an open circle, $\bigcirc$. The Hall-Petch-like relationship, which is indicated by the approximated line in the figure, was found.

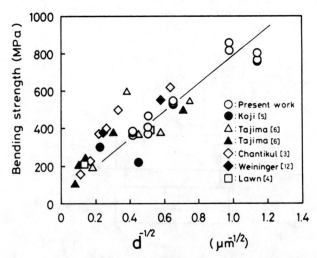

Figure 2. Relationship between bending strength and inverse of square root of grain diameter, $d^{-1/2}$, in alumina ceramics.

Fractographs of the bending specimens are shown in Fig.3. On the basis of the fractographic observations, it was difficult to identify the location and size of the fracture origin, but some small surface and internal defects were found. Quasi-cleavage facets of from 4 to 8 grains in size were dominant on the fracture surface regardless of grain size, as shown in Fig.3(b).

Fracture toughness
From the results of fracture toughness tests, the relationship between fracture toughness and grain diameter is shown in Fig.4. The fracture toughness evaluated by the CSF method was

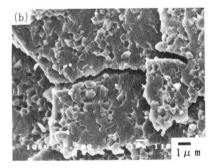

500 μm   1 μm

Figure 3. Fractographs of bending specimens. (a) Macroscopic view of the fracture surface of the specimen with a grain size of 2.4 μm, (b) Microscopic fracture surface of the specimen with a grain size of 0.8 μm.

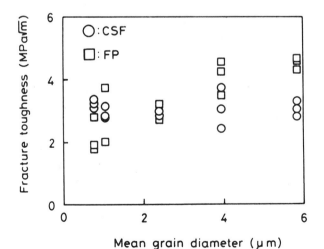

Figure 4. Relationship between fracture toughness and grain diameter.

almost independent of grain size, while that obtained by the FP method increased with increasing grain size in the grain size region larger than 3 μ m.

Fractographs of the CSF specimens are shown in Fig.5. In these, a ring-shaped crack region at the tip of surface crack was observed, which was significant for specimens with small grain sizes. Intergranular fracture was dominant in the ring-shaped crack region, while transgranular fracture was dominant in the indentation surface crack region and the unstable fracture region, where quasi-cleavage fracture facets of from 4 to 8 grains in size were observed. Intergranular fracture could be induced by stress corrosion cracking during the indentation cracking or fracture toughness test. The fracture toughness values shown in Fig.4 were evaluated by using the crack length including the ring-shaped region.

Fractographs of the FP specimens are shown in Fig.6. From the fractographic observations it was found that intergranular fracture was dominant in the fatigue precrack region. On the other hand, transgranular fracture was dominant in the unstable fracture region, where quasi-cleavage fracture facets of from 4 to 8 grains in size were observed.

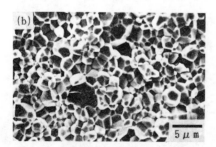
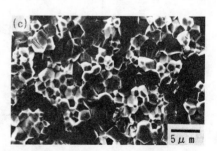

Figure 5. Fractographs of the CSF specimen with a grain size of 2.4 μ m. (a) Overview of a surface crack, (b) SCC ring crack region, (c) Unstable fracture region.

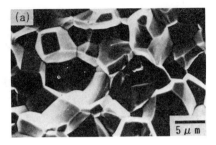

Figure 6. Fractographs of the FP specimen with a grain size
of 5.9 μ m. (a) Fatigue precracked region,
(b) Unstable fracture region.

Static fatigue crack growth rate

Static fatigue crack growth curves for the alumina specimens
tested are shown in Fig.7. The crack growth rates of the
specimens with small grain diameters of 0.8 and 2.4 μ m were
almost identical and higher than those of the specimen with
larger grain diameters of about 5.9 μ m.

From the fractographic observations, intergranular
fracture was found to be dominant regardless of grain size, as
shown in Fig.8.

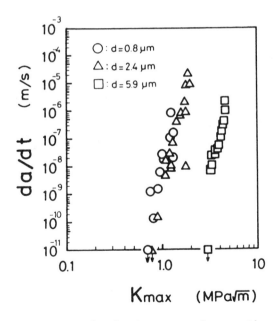

Figure 7. Static fatigue crack growth curves.

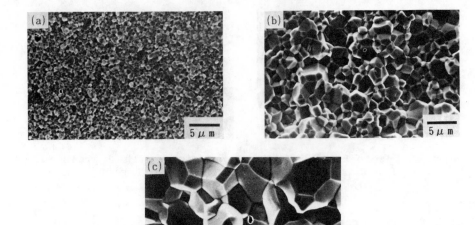

Figure 8. Fractographs of the static fatigue crack growth
specimens. (a) For grain sizes of 0.8 μ m,
(b) 2.4 μ m, (c) 5.9 μ m.

DISCUSSION

Effect of grain size on bending strength

The bending strength of the alumina ceramics tested in the
present study increased with decreasing grain size, as shown in
Fig.2. Since the fracture toughness value of the present
material tested is almost constant and independent of grain
size as discussed below, the expected size of the initial
surface defect, 2a, can be calculated by substituting the
constant fracture toughness value and the bending strength into
the Newman-Raju equation [8]. The estimated defect size, 2a, is
almost directly proportional to the grain diameter, d, and the
mean value of, 2a, is about 7.5 times the grain diameter, as
shown in Table 2. On the basis of the fractographic
observations, the diameter of quasi-cleavage fracture facets,
which were the dominant mode of fracture in bending specimen,
is 4 to 8 times the grain diameter. The formation of a quasi-
cleavage fracture facet over several grains suggests that the
grain boundary does not play an important role in contrast to
the case with metallic materials, where the grain boundary
obstructs the movement of dislocation and slip. Therefore, it
is suggested that the apparent dependency of the bending
strength on grain size results from the pre-existence of
defects. The size of such defects is in proportion to the grain
size or to the size of quasi-cleavage fracture facets, which
are also proportional to the grain size. This apparent
dependency could also be a consequence of the formation, during

TABLE 2
Expected diameter of initial surface crack 2a and mean
grain diameter d .

Specimen	d (μ m)	2a (μ m)	2a/d
A	0.8	7.3	9.5
B	1.1	7.5	7.1
C	2.4	17.9	7.5
D	3.9	26.9	6.8
E	5.9	37.7	6.4

the bending test, of a starting surface crack with the same
size of quasi-cleavage fracture facets, which is also
proportional to the grain size.

Several experimental results on the bending strength of
alumina specimens with various grain sizes have been reported
[2]~ [6]. These results are in close agreement with the present
result, as shown in Fig.2.

Effect of grain size on fracture toughness
The fracture toughness evaluated by the CSF method was almost
constant regardless of grain size, while that by the FP method
increased with increasing grain size in the larger grain size
region. Recently, Suresh et al. [10] reported that the fracture
toughness of an alumina specimen with a through-the-thickness
fatigue-precrack increased with increasing grain size. Vekinis
et al. [11] also reported a grain size dependency of fracture
toughness similar to the present result for the fatigue
precracked specimen. They believed that the higher fracture
toughness of the specimen with larger grain sizes resulted from
the higher crack growth resistance (R-curve behavior
[11]~ [14]) during slow crack growth prior to unstable fracture
in the specimen with larger grain sizes. Their explanation for
the higher crack growth resistance in the specimen with larger
grain sizes is that the main mechanism of R-curve behavior in
alumina ceramics is considered to be "grain-bridging"
[11],[13],[14]. The larger grains result in larger bridging
length and consequently in higher crack growth resistance.
Figure 9 shows an example of "grain-bridging" in the slow crack
growth region prior to unstable fracture in the fatigue
precracked specimen. However, since no evidence of grain-
bridging is found in the unstable fracture region, this
apparent toughening due to R-curve behavior does not mean true
toughening of the material.

Figure 10 shows the increase in stress intensity factor,
K, with crack extension for the FP and CSF specimens. It is
found that the K-value of the CSF specimen increases with crack
extension more rapidly than in the FP specimen. The slow crack
growth prior to unstable fracture, which was found in the
through-the-thickness FP specimen, does not occur in the CSF
specimen.

Figure 9. "Grain bridging" in the slow crack growth region observed from the direction parallel to the fracture surface of the specimen with a grain size of 5.9 μ m.

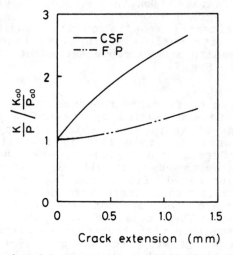

Figure 10. Relationship between the K-value and crack extension for the FP and CSF specimens.

From the above discussion, the fracture toughness evaluated by the FP method should be understood to be the apparent value due to the existence of stable crack growth, where the fracture toughness values depend on the configurations and dimensions of the specimens. On the other hand, the fracture toughness evaluated by the CSF method gives the validated value according to the standard for plain-strain fracture toughness, which represents the inherent characteristic for fracture resistance of the material, independent of specimen geometry and size. Consequently, it can be concluded that the fracture toughness of alumina ceramics is independent of grain size within the range of grain sizes discussed here.

Effect of grain size on static fatigue crack growth rate
The static fatigue crack growth rate of alumina ceramics is reduced with increasing grain size, as shown in Fig.7. If the SCC characteristics at the grain boundary are independent of grain size then the main reason for the lower crack growth rate in the specimen with larger grain sizes will be that the larger grains result in a larger bridge length and consequently induce increased crack growth resistance. Further research is needed of the effect of grain size on the SCC property at grain boundaries.

CONCLUSIONS

Bending strength, fracture toughness, and static fatigue crack growth tests of alumina specimens with various grain diameters ranging from 0.8 to 5.9 μ m were carried out. The main results are summarized as follows.
(1) The bending strength of alumina ceramics increases with decreasing grain size. The apparent grain size dependency of the bending strength is induced by pre-existing defects. The size of these defects is in proportion to the grain size.
(2) The validated value of the fracture toughness in alumina ceramics is independent of grain size. The apparent fracture toughness value obtained by specimens with slow crack growth increases with increasing grain size due to R-curve behavior.
(3) The static fatigue crack growth rate of alumina ceramics is reduced with increasing grain size.

REFERENCES

1.Barrett,C.R., Nix, W.D. and Tetelman,A.S., The principles of engineering materials. Prentice-Hall Inc., 1973,80.

2.Ting, Jyh-Ming, Lin, Y., Ray and Ko, Ying-Hsiang, Effect of powder characteristics on microstructure and strength of sintered alumina. Ceram. Bull., 1991, 70, 1197-1172.

3.Chantikul, P., Bennison, S.J. and Lawn, B.R., Rolle of grain size in the strength and R-curve properties of alumina. J. Am. Ceram. Soc., 1990, 73, 2419-2427.

4.Lawn,B.R., Freiman, S.W. and Baker, T.L., Study of microstructural effects in the strength of alumina using controlled flaws. J. Am. Ceram. Soc., 1984,71,C-67-C-68.

5.Koji,T., Wakayama, S. and Nishimura, H., Influence of strain rate and grain size on the critical microcracking stress in alumina. Reprint of JSME spring Ann. Mtg.,900-14,1990,159-161.

6.Tajima, Y. and Urasima, K., Improvement of strength, fracture toughness and reliability of engineering ceramics. Yogyo-kyokai-shi, 1990, 25, 96-100.

7.Mutoh,Y., Tanaka,k. and Miyahara,N., A method for evaluating the fracture toughness of ceramics. Trans. JSME, 1989, 55A, 2144-2151.

8.Newman,Jr.,J.C. and Raju,I.S., An empirical stress-intensity factor equation for the sureface crack. Eng. Fract. Mech., 1981,15,185-192.

9.Sadahiro, T. and Takatsu, S., A new precracking method for fracture toughness testing of cemented carbides. Modern Develop. in Powder Met., 1981,12,561-572.

10.Suresh,S., Ewart,L., Maden,M., Slaughter,W.S.and Nguyen,M., Fracture toughness measurements in ceramics: Precracking in cyclic compression. J. Mat. Sci.,1987,22,1271-1276.

11.Vekinis,G., Ashby,M.F. and Beaumont, P.W.R., R-curve behavior of Al_2O_3 ceramics. Acta Metall. Mater.,1990,38, 1151-1162.

12.Wieninger,H. and Kromp,K., Crack resistance curves of alumina at high temperature. J. Mat. Sci., 1987,22,1352-1358.

13.Steinbrech,R.W., Reichl,A. and Schaarwächter,W., R-curve behavior of long cracks in alumina. J. Am. Ceram. Soc., 1990,73,2009-2015.

14.Swanson,P.L., Fairbanks, C.J.,Lawn, B.R., Mai, Y.M. and Hockey, B.J., Crack-interface grain bridging as a fracture resistance mechanism in ceramics: I experimental study on alumina. J. Am. Ceram. Soc.,1987,70,279-289.

CRACK PROPAGATION IN ALUMINA DURING IN-SITU STRAINING BY TEM

Y.IKUHARA, T.SUZUKI, M.KUSUNOKI AND Y.KUBO
Japan Fine Ceramics Center, Nagoya, Japan, 456

ABSTRACT

Crack propagation in sintered alumina was observed by an in-situ straining experiment in a 400 kV transmission electron microscope (TEM) at room temperature and 1070 K. Crack deflection and healing at grain boundaries were observed at 1070 K. Dislocations were generated at crack tips in grains at both temperatures. The strain contour line around the crack tips periodically oscillated during crack propagation, which was explained by the generation of periodic dislocations.

INTRODUCTION

To understand fracture phenomena in ceramics, it is very important to clarify the internal structure of crack tips. For observing the internal structure, TEM is the most useful method. To prepare the specimen for TEM observation, cracks are generally introduced ex-situ by an indentation method, then the specimen is thinned by argon ion bombardment. In this case, it is possible that the structure of the crack tips changes because of some artifact such as dislocation escape due to the thin film effect [1] or crack closure due to residual stress [2].

In contrast, an in-situ TEM method for observing fracture behavior is very effective for clarifying the microstructure of a crack tip, because the crack can be dynamically produced in the TEM to eliminate the artifact mentioned above. Kobayashi and Ohr [3] succeeded in observing crack propagation behavior in molybdenum and in proving the existence of a dislocation free zone. Recently, Chiao and Clarke [4] reported a dislocation emission from crack tips in silicon by an in-situ TEM method. In the case of relatively ductile ceramics such as MgO [5] and ZrO_2 [6], TEM in-situ straining experiments have already been accomplished obtaining very valuable information. There are, however, no reports of experimental studies of the crack tip during fracture, with the more brittle ceramics such as Al_2O_3, Si_3N_4, and SiC. This is because the speed of crack propagation is too fast to observe by TEM.

Recently, we have developed a CIT (Crack Induced Tension) method [7] that now makes it possible to observe crack propagation behavior under stable conditions using in-situ TEM. We believe this technique can overcome the above mentioned problems.

This study reports the investigation of crack propagation behavior in sintered alumina by in-situ strain experiments in a TEM.

MATERIALS AND METHODS

Sintered alumina (99.7 % pure JFCC, Referceram) with an average grain size of 2 μm was sliced, and thinned by argon ion bombardment to prepare thin foils for TEM. The electron microscope used was a side-entry JEOL 4000FX. In-situ observations were carried out both at room temperature and at 1070 K at a tensile speed of about 1×10^{-5} mm/s by CIT [7].

RESULTS AND DISCUSSION

Figure 1 shows a moment of intergranular fracture. When stress was applied at a fairly high speed of 10^{-3} mm/s, fracture occurred through a grain boundary instantaneously.

Figure 1. Fracture through a grain boundary in a thin film of sintered alumina at room temperature subjected to a tensile speed of 1×10^{-3} mm/s.

Fig.2 is a series of micrographs of intergranular fracture at 1070 K at a tensile speed of 1 x 10^{-5} mm/s resulting from application of the in-situ CIT method. The area shown is at a grain boundary triple junction. Fracture develops from (a) to (c). The initial microcrack along the grain boundary can be seen in (a). As the tensile stress is applied, a fan-shaped strain contrast appears and its intensity becomes stronger as shown in (b). At (c), a crack is deflected at the grain boundary and the strain contour changes. These micrographs directly prove the existence of strain at the intersection of cracks and grain boundaries, along with crack deflection by the grain boundary.

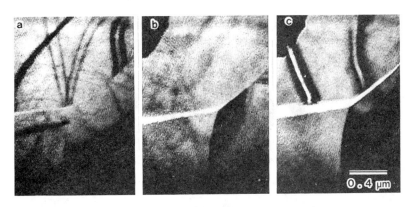

Figure 2. A series of TEM micrographs obtained by in-situ experiment at 1070 K, showing intergranular fracture through the grain boundary at a triple junction. Crack deflection is clearly seen after the appearance of a fan-shaped strain contour at the intersection of the microcrack and the grain boundary.

When the applied stress was removed, crack healing was observed at the grain boundary as shown in Fig.3. The initial microcrack through the grain boundary is seen in (a). The crack opening displacement increases with increasing stress as shown in (b), and the crack completely closed as soon as the applied stress was removed. This healing phenomenon occurs readily with grain boundary fracture at high temperature.

Fig.4 is a series of micrographs of transgranular fracture at 1070 K. The development of a crack is shown from (a) to (c). The arrow indicates the crack tip in each micrograph. The wide black fringe (indicated by the triangle) around the crack tip in the border corresponds to the contour line at constant strain. As the crack develops, this fringe oscillates periodically, that is, the diameter of the contour region becomes alternately smaller and larger as shown in figure 4.

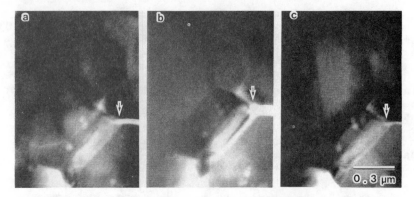

Figure 3. A series of TEM micrographs, showing crack healing at a grain boundary. The crack was completely closed when the applied stress was removed.

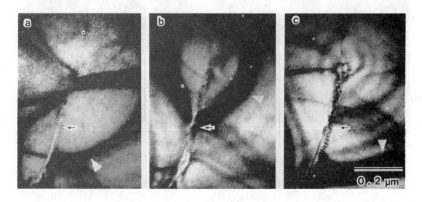

Figure 4. A series of TEM micrographs obtained by an in-situ CIT experiment at 1070 K. The sequence shows the development of a crack, the accompanying change in strain contour around the crack tip, and the generation of a dislocation at the crack tip. The crack tip and the strain contour are indicated by the arrow and the triangular mark respectively in each micrograph.

Fig.5 is a magnified micrograph of the crack tip. In the vicinity of the crack tip, an array of dislocations is clearly observed over a width of 0.25 μm. In addition to the dislocations array, the dislocation wake along the crack walls is also observed over a range of 0.6 μm. Beyond the wake, dislocations disappear probably because of thin foil effects [1]. A dislocation free zone [8] could not be found.

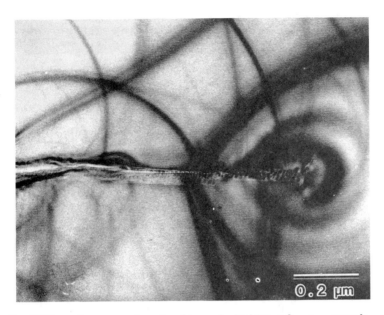

Figure 5. TEM micrograph in the vicinity of a crack propagating at 1070 K, indicating an array of dislocations at the tip and a dislocation wake along the crack walls.

Fig.6, a schematic plot of stress vs. distance from the crack tip, helps to explain this phenomenon. Line A shows the stress distribution in front of a crack tip when no dislocations are emitted. When dislocations are emitted from the crack tip, the shielding of the latter by the dislocations changes the stress distribution to line B. A strain contour corresponding to a stress σ_c is present at a distance r_1 in case B, and r_2 in case A. Then the region of the contour oscillation occurs with periodic generation of dislocations over the distance between r_1 and r_2. Maeda and Fujita [9] proposed a dynamic work hardening model at the crack tip in covalent

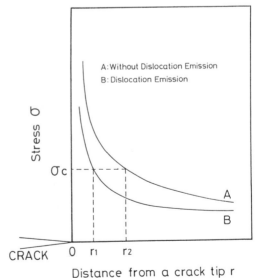

Figure 6. A schematic plot of stress vs. distance from the crack tip.

ceramics. This model, which describes the alternate occurrence of dislocation emission and its shielding effect, corresponds to the observations in the present study of contour line oscillation.

Fig.7 shows an example of crack tip microstructure during the application of tensile stress at room temperature. In this case, the strain contour around the crack tip also oscillates during crack propagation, suggesting that the dislocations are emitted from the crack tip in the figure and depend on the crystal orientation, i.e. the inclination of the slip plane to the plane of the thin film.

Hockey and Lawn [10] report that are no dislocations generated from crack tips, and thus concluded that materials such as Al_2O_3 and SiC are essentially brittle. Actually, we also did not find dislocations in the present specimens when cracks were introduced by an indentation method. Therefore, if TEM observations are conducted on cracks that are introduced, there is a possibility that the structure of the crack tip changes during the specimen preparation, i.e. during ion-thinning. If the specimen is sufficiently thin for HREM observation, some dislocations are likely to escape from the thin film because of thin film effects. What determines the brittleness of a material is considered to be the size of the process zone. Ceramics, such as sintered alumina, may be brittle because the size of the microplastic zone is $\leq$ 0.3 µm. Although the thin film effects and the planar stress condition must be taken into account, we believe that these in-situ experiments provide an effective method to observe the internal structure of crack tips in thin films.

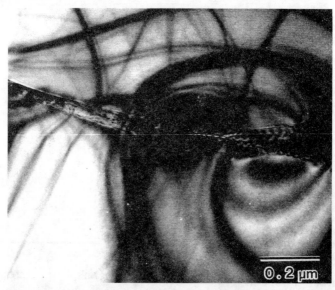

Figure 7. TEM micrograph of a crack propagating through a grain at room temperature, indicating the presence of round strain contours and radiated dislocations in front of the crack tip.

CONCLUSIONS

Crack propagation in sintered alumina was observed by in-situ strain experiments in a 400 kV transmission electron microscope (TEM) at room temperature and 1070 K. The following observations were made;

1. Fracture at a high strain rate occurred through a grain boundary at 1070 K.

2. Crack deflection through a grain boundary was clearly observed. A fan shaped strain appeared at the intersection of the crack and the grain boundary.

3. Crack healing at a grain boundary was observed at 1070 K. The crack was completely closed when the applied stress was removed.

4. Dislocations were generated at the crack tip both at room temperature and at 1070 K in the case of transgranuler fracture. The size of the dislocation arrays were $0.2 \sim 0.3$ μm.

5. The size of the stress contour line around the crack tips oscillated periodically during crack propagation. This can be explained by the periodic generation of dislocations.

ACKNOWLEDGEMENTS

The authors would like to thank the members of JFCC machining group for their help in sample preparation, Dr. H. Awaji of JFCC, as well as Dr. K. Higashida of Kyoto Univ. for useful discussions, and the members of JEOL Ltd. for assistance with technical matters.

REFERENCES

1. Mader, S., Seeger, A,. and Thieringer, H.M., Work Hardening and Dislocation Arrangement of fcc Single Crystals. Appl. Phys., 1963, **34**, 3376.

2. Lawn, B.R., Hockey, B.J., and Widerhorn, S.M., Atomically sharp cracks in brittle solids : an electron microscopy study. J. Mat. Sci., 1980, **19**, 1207-1223.

3. Kobayashi, S., and Ohr, S.M., In situ fracture experiments in b.c.c. metals. Proc. the 37th Ann. Meeting of EMSA, ed. G.W.Bailey, 1979, pp. 424.

4. Chiao, Y.H., and Clarke, D.R., Direct observation of dislocation emission from crack tips in silicon at high temperatures. Acta Met. 1989, **37**, 203-219.

5. Appel, F., Messerschmidt, U., and Kuna, M., Crack Propagation in MgO during In-Situ Deformation in High-Voltage Electron Microscope. Phys. Stat. Sol. (a) 1979, **55**, 529-536.

6. Rühle, M., and Waidelich, D., In-Situ Observations of martensic transformation of Enclosed-ZrO$_2$ Particles. The Proceedings of International Symposium on In Situ Experiments with HVEM Osaka University, 1985, pp.157-167.

7. Ikuhara, Y., Suzuki,T., and Kubo, Y., The development of CIT and its applications. to be published in JFCC Review.

8. Majumdar, B.S., and Burns, S.J., Crack Tip Dislocations in MgO. Scripta Metall., 1980, **14**, 653-656.

9. Maeda, K., and Fujita, S., Microscopic Mechanism of Britrle-Ductile Transition in Covalent Ceramics. Jap. J. Appl. Phys., Series 2 (Lattice Defects in Ceramics), 1989, pp.25-31.

10. Hockey, B.J., and Lawn, B.R., Electron Microscopy of Microcracking about Indentations in Aluminum Oxide and Silicon Carbide. J. Mat. Sci., 1975, **10**, 1275-84.

INFLUENCE OF ENVIRONMENT ON STRENGTH OF CERAMICS

T. S. Sudarshan[1] and T. S. Srivatsan[2]
1. Materials Modification Inc.
2929-P1, Eskridge Center
Fairfax, Virginia 22031

2. Department of Mechanical Engineering
The University of Akron
Akron, Ohio 44325-3903

ABSTRACT

The increasing need for engineering components in both civilian and military applications to operate at elevated temperatures for prolonged periods of time has led to the development and use of novel ceramic materials both in the monolithic form and as composite constituents. Many of these applications involve exposure to environments containing hydrogen, water vapor and other aggressive species. In this paper, the effects of hydrogen-containing environment on the strength and mechanical response of monolithic ceramics and ceramic-metal interfaces is examined. The mechanisms contributing to strength degradation are detailed and the response of interfaces to the environment is highlighted.

INTRODUCTION

The increasing demand of the aerospace industry for high temperature, high strength materials has led to the development of ceramic-matrix composites [CMCs], metal-matrix composites [MMCs] and intermetallic-matrix composites [IMCs] [1-4]. This demand has been accentuated by the desire to create engines that can operate at elevated temperatures for prolonged periods of time, and are also able to achieve the goal of single stage to orbit [SSTO]. Using a fiber-reinforced architecture resulted in a material having several advantages over the monolithic counterpart and even multiphase materials, but with limited microstructural control. Some of the advantages include, but may not be limited to, improved elevated temperature strength, high modulus, low density, adequate fracture toughness, creep resistance, oxidation resistance, improved fatigue resistance, and increased density normalized properties that follow from lower net material density [2,3]. The potential payoffs in using these newer generation materials include reduced cooling requirements, reduced weight of the aircraft engine and increased thermodynamic efficiencies which will lead to reduced fuel consumption [4]. However, much of these benefits do not come without problems that must be addressed. For short term service requirements, high temperature excursions, thermomechanical compatibility and environmental resistance are key issues that must be examined and addressed. For example, oxidation and embrittlement play an important role in determining the stability of composites. In fact, oxidation at elevated temperatures significantly affects the microstructure of the reaction zone, matrix and outer fiber coating of the composites [5,6]. For long term high temperature applications, thermally activated processes, such as interdiffusion and interfacial reactions between the reinforcing fibers and the matrix must be well understood.

In developing composite systems for high temperature applications, the use of ceramics is increasing steadily largely because of the improved mechanical integrity afforded by adequate toughness and by process control. Current theories suggest that microcracking [7], crack deflection [8] and crack bridging [9,10] are the predominant toughening mechanisms, although the relative contributions of each mechanism to toughening have not been clearly delineated. However, there is general agreement that the interfaces between the ceramic reinforcements, in the form of whiskers, and the matrix are of prime importance in enhancing toughness. Interfaces that are weakly bonded tend to enhance the toughness of composites by facilitating crack bridging and whisker pullout during crack propagation, while interfaces that are strongly bonded facilitate whisker fracture which yields little or no increase in toughness [11]. A necessary and essential technical prerequisite for adequate mechanical integrity is the fracture resistance of the interface. Interfaces in metal or metallic alloy bonded to a ceramic must either sustain mechanical forces without failure or exhibit controlled debonding. Consequently, interfaces exert an important influence in determining physical processes and in technological applications such as metal-matrix composites [MMCs], catalyst supports, adhesion of films and protective coatings, and microelectronics packaging systems [12-14]. Interfaces also play an important role in internal and external oxidation or reduction of materials. Overall composite properties are dominated by the interface such that bounds must be placed on the interface debond and sliding resistance in order to have a composite with attractive mechanical properties. For the joining of ceramics, bonds having strengths well in excess of the strength of the ceramic can be achieved using a ductile metal. The strength and fracture energy of the resulting ceramic metal bond is a function of several competing variables such as [15]:

(a) the thermal and elastic mismatch across the interface,
(b) the relative thickness of the metal layer,
(c) plastic flow in the metal,
(d) morphology of the interface, and
(e) the presence of defects or interfacial compounds at or near the interface.

In recent years, several elegant theoretical and experimental studies have been conducted to understand and document the fracture toughness of metal/ceramic interfaces [16,17], and on developing a relationship between interface strength and fracture toughness of metal-ceramic composites [18,19]. However, few studies have focussed exclusively on understanding the importance of interfaces during oxidation and or exposure of material to aggressive environments, particularly, those containing water vapor and hydrogen. This may be deemed to be important as failure of the entire structural component can result at stresses far below those required for catastrophic failure. Accordingly, the intent of this paper is to examine the possible role of interfaces on strength and environment susceptibility of materials, particularly ceramics. We begin with a brief discussion on adhesion and strength characteristics of interfaces and subsequently examine environment influences on the behavior of certain monolithic ceramics and ceramic joints.

ADHESION AND STRENGTH OF INTERFACE

The driving force for the formation of a metal/ceramic interface is the yield in energy when intimate contact is established between the metal and ceramic surfaces [20]. To have a high rate of interaction, the surfaces have to be brought into excited states. Therefore, both temperature and atmosphere are important variables, as well as the properties and structures of the surfaces.

The physical interaction between a metal and a ceramic is the work of adhesion, W_{ad}. When perfectly clean, defect-free surfaces are brought into contact, energy is released in accordance with the relation:

$$W_{ad} = \gamma_C + \gamma_M - \gamma_{MC} \tag{1}$$

where γ_C and γ_M are free energies of the relaxed surfaces of the ceramic and metal respectively, γ_{MC} is the energy of the relaxed interface between the metal and the ceramic. W_{ad} is the reversible work released per unit area of interface formed by two free surfaces. However, direct measurement of W_{ad} is not possible [21]. It is determined by measuring the contact angle, say θ, established by a solid metal in contact with a ceramic. Alloying additions have a profound influence on the thermodynamic stability of an interface. They tend to segregate at the interface by Gibbsian absorption [20]. An example is the segregation of chromium at various metal/Al_2O_3 interfaces. This results in a rearrangement of the interface into a more relaxed and stable structure with a lower interfacial energy, and thus a lower work of adhesion [20]. Such segregant effects are a major issue in metal-ceramic bonded couples and joints [22,23]. However, trace element impurities at low concentrations, well below one percent by weight and usually in the range of hundreds of ppm or less, segregate to interfaces resulting in a locally high concentration at the interfaces which tends to weaken them and promotes reactions during exposure to aggressive environments.

Strength of the interface can significantly change the local stress-displacement curve for deformation in the metal, thereby having a marked influence on toughness of the composite material. The properties of metal-ceramic interfaces are also likely to change when exposed to high temperatures either during processing or in service. Thus, attempts to use thin interfacial coatings of foreign materials to control properties may not be reliable unless accurate prediction of the reaction rate and an understanding of the influence of these reactions on mechanical properties is well understood. Predictions of interfacial properties based upon heat treatments of interfaces that move them to their equilibrium state is more reliable than those based on coatings that are far from thermodynamic equilibrium.

Shieu and coworkers [24] examined a NiO-Pt interface and analyzed the influence of reactions occurring at the interface on both structure and shear strength. The interfacial reactions were stimulated by suitable choice of annealing temperature, time and oxygen partial pressure. The thickness of the intermetallic compounds [NiPt] that formed was carefully monitored. The ultimate shear strength of the NiO-Pt interface with and without different interlayers was determined using the periodic cracking method [25]. Application of load to the Pt plate resulted in the development of shear stresses in the immediate vicinity of the interface due to differences in axial displacement between the NiO and Pt, as shown in Figure 1. As the strain in the Pt plate increases, periodic cracks develop in the thin NiO film. The cracks divide the film into many parallel strips. When the cracking ceases, the maximum tensile stress which occurs at the center of the strip is equal to the tensile strength of the NiO film. The tensile stress is assumed to be uniform through the thickness of the NiO film. Since the NiO single crystal is brittle it is assumed to deform elastically and the fracture stress is estimated using Hooke's Law using the Young's Modulus of NiO and the fracture strain obtained from the crack density versus strain curve [24]. Assuming the normal tensile stress at the interface near the edge of the strip does not cause delamination, the tensile stress which can cause fracture in the NiO film is related to the shear stress at or near the interface through the integral equation:

$$\sigma(x) = \frac{1}{\delta} \int_{0}^{x} \tau(x)\, dx \qquad (2)$$

where σ is the tensile stress in the film, δ is the film thickness, $\tau(x)$ is the shear stress at the NiO-Pt interface and x is the coordinate along the interface with origin at the crack. The maximum shear stress, τ_{MAX}, at the interface is therefore:

$$\tau_{MAX} = \frac{K\delta\sigma_f}{\lambda_{MAX}} \qquad (3)$$

where λ_{MAX} is the maximum crack spacing in the film and K is a constant that depends on the shear stress distribution along the interface. Figure 2 shows the distribution of shear stress and the resultant tensile stress in the ceramic film of a metal-ceramic joint. Furthermore, the differences in elastic properties across the interface of two dissimilar materials induces a shear component [26]. For plane strain, the elastic moduli mismatch is obtained using parameters of Dundurs [27].

$$\alpha = \frac{E_1' - E_2'}{E_1' + E_2'} \qquad (4)$$

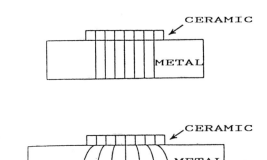

Figure 1. Schematic showing load transfer between the metal and ceramic
via. shear lag of axial displacement in the loading direction
(Ref. 24).

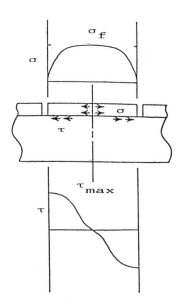

Figure 2. Shear stress distribution along the interface between the
metal and the ceramic, and the resultant tensile stress in
the ceramic strip (Ref. 24)

$$\beta = \frac{1}{2} \left\{ \frac{\mu_1(1-2\nu_2) - \mu_2(1-2\nu_1)}{\mu_1(1-\nu_2) + \mu_2(1-\nu_1)} \right\} \tag{5}$$

where $E' = E/1-\nu^2$ and μ_1 and μ_2 are shear modulus, and ν_1 and ν_2 is the poisson's ratio of materials 1 (ceramic) and 2 (metal) across the interface, and the parameters change signs when the materials are switched.

INFLUENCE OF ENVIRONMENT ON STRENGTH

Mechanical properties such as strength and toughness of metal/ceramic joints or ceramic/ceramic joints having metal layers are not only dependent on the chemistry of interfacial bonding on an atomic scale but also on macroscopic factors such as thermal stress, interfacial configurations, plasticity of metals and the presence of defects and cracks in reaction layers. In fact, the mechanical properties of most ceramics are sensitive to the presence of water in the environment, which accelerates the stress corrosion phenomenon. From a technological standpoint, the stress corrosion problem becomes very important when materials are to be used as parts of structural components which find exposure not only to air but to repeated cycling in hot combustion gases containing water vapor and hydrogen gas.   Thus, the elevated temperature compatibility and hydrogen permeability of ceramic materials are critical factors in enhancing their selection for use in monolithic form and as composite constituents in metal matrix composites [MMCs], ceramic matrix composites [CMCs] and even carbon/carbon [C/C]. For example, aluminide matrix composites form scales of alumina (Al_2O_3) in an oxygen-containing environment, and alumina is a promising fiber candidate for these composites [28].   In this section, we will consider the environmental response of pure ceramic materials such as monolithic mullite, silicon carbide [SiC], silicon nitride and alumina, and even metal-ceramic joints.   In particular, the influence of environment on strength degradation and mechanical performance will be examined.

Monolithic Mullite [$3Al_2O_3$-$2 SiO_2$].

This material due to its high strength and low thermal expansion coefficient is an ideal candidate for use as a pressure vessel in the hot zone of a ceramic stirling engine. The use of mullite in the stirling engine requires that the material retain much of its strength even after prolonged exposure to hydrogen-containing environments at elevated temperatures. At elevated temperatures, corrosion of the mullite is promoted as a result of material loss and/or selective pitting which degrades fracture strength.

Weight loss was detected for mullite exposed to hydrogen at temperatures as high as 1425°C [29]. The loss in weight was dependent on hydrogen flow rate and the reaction between gaseous hydrogen and mullite was diffusion controlled [29]. An increased weight loss resulted in degradation of compressive strength. The mullite-hydrogen gas reaction was process dependent on the transportation of the gaseous species through the porous product layer and also through the gaseous boundary layer [30]. The diffusion of hydrogen through the porous layer was not the rate limiting step, rather it was the reduction rate. The reaction of mullite with hydrogen resulted in a loss of silica leaving behind a porous product layer of alpha alumina on the surface [30].

Herbell and coworkers [28] evaluated the corrosion behavior of near stoichiometric mullite [$3Al_2O_3$-$2SiO_2$] in pure hydrogen gas at temperatures of 1050°C and 1250°C for times up to 500 hours. At the end of hydrogen exposure the samples were removed from the furnace, weighed, and measured. The exposed specimens were then stressed to failure in a four point bend fixture. These researchers found the corrosion of mullite to be severe at 1250°C. The enhanced corrosion was manifested by the formation and deposition of silica whiskers near the outlet end of the furnace tube. The formation of SiO and H_2O gases results from the reaction of SiO_2 with hydrogen

$$SiO_2 \text{ (s)} + H_2 \text{ (g)} = SiO \text{ (g)} + H_2O \text{ (g)}$$

The SiO is removed from the reaction layer by the bulk gas and transported to the cooler outer end of the furnace tube where conditions favor its reoxidation to SiO_2. Removal of SiO_2 from the glassy phase and the mullite grains leaves behind regions of porous alpha alumina. The alumina/silica [Al/Si] ratio increased with increased exposure to the gas at 1250° C (Figure 3). The removal of SiO_2 was found to degrade the density and increase the porosity near the surface [28]. The observed decrease in density was found to be linear with time up to the maximum of 500 hours (Figure 4). The porous surface layer resulting from hydrogen corrosion is shown in Figure 5. The depth of penetration of porosity was found to be more severe after 500 hour exposure at 1250°C than at 1050°C.

Exposure to hydrogen at the elevated temperature was also found to have an influence on fracture strength (Figure 6). At 1050°C, exposure to the hydrogen environment produced appreciable loss in strength for all time levels. The loss in strength being noticeable in the initial (25-50) hours of exposure and reaching 22 percent after 500 hour in hydrogen. The observed loss in strength of mullite after prolonged exposure (500 hours) to the elevated temperature was attributed primarily to the formation of a thin porous reaction layer due to corrosion of the glassy grain boundary phase between the mullite grains [28]. Transmission electron microscopy observations revealed the presence of a faulted crystalline grain boundary phase in samples exposed to 1050°C. Strength reductions of up to 28 percent were observed in the presence of this phase. However, no such phase was detected in samples exposed to 1250°C. The rapid increase in strength at short exposure times, at 1250°C, was rationalized as being due to healing of the surface cracks by the flow of SiO_2 on the surface. The combined influence of environment and temperature revealed that pure mullite without the presence of a secondary glass phase may be suitable for use in hydrogen-containing environments at temperatures below 1250°C [28].

Post exposure test bars of mullite exposed to an environment of wet hydrogen for 100 hours at temperature of 1100°C revealed a progressive loss of strength (Figure 7) with a maximum loss being 14 percent. The microstructure of the exposed material showed a needlelike structure of the mullite grains (Figure 8). This was attributed to removal of the silica-rich intergranular glassy phase in the high temperature hydrogen environment [31].

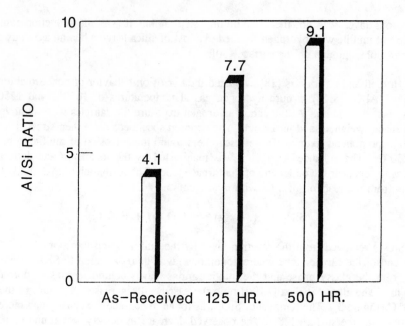

Figure 3.   Ratio of Al/Si on the surface of mullite exposed to hydrogen
at 1250°C (Ref. 28).

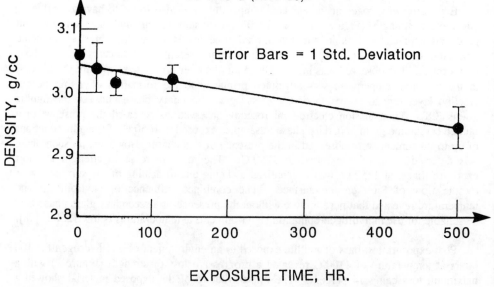

Figure 4.   Variation of density of mullite as a function of exposure time
to hydrogen environment at 1250°C (Ref. 28).

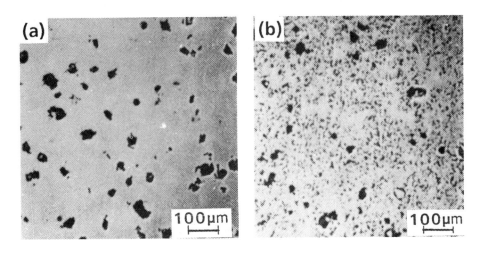

AS RECEIVED   500 HR. 1250°C

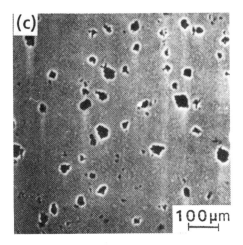

500 HR 1050°C

Figure 5.   Micrographs showing the reaction layer after exposure to hydrogen
at 1250°C (Ref. 28).

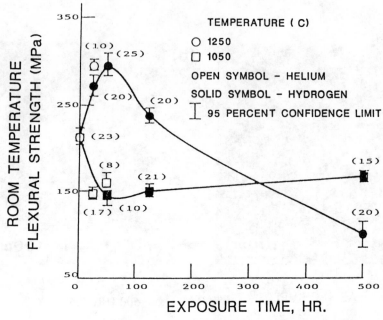

Figure 6. Variation of the fracture strength of mullite as a function of
exposure time. Number in parentheses indicates sample sizes
(Ref. 28).

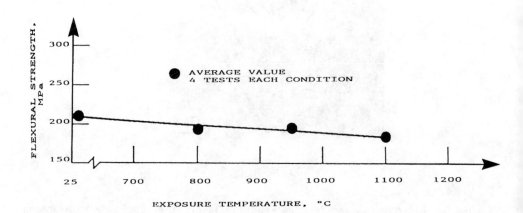

Figure 7. Influence of wet hydrogen, at elevated temperatures, on room
temperature strength of mullite (Ref. 31).

Aluminum Oxide (Al_2O_3)

The thermodynamic stability of aluminum oxide in hydrogen was studied by Herbell and coworkers [31]. The stability of Al_2O_3 was found to decrease with increased pressure at high temperatures, and at 1 atmosphere pressure the ceramic found to be stable in pure hydrogen. Based on equilibrium partial pressure, this material was expected to be stable in both hydrogen, and hydrogen and water vapor-containing environments at elevated temperatures. However, the room temperature flexural strength decreased by about 20 percent following exposure to wet hydrogen for 100 hour at 1400°C. The degradation in strength was attributed to changes in intrinsic microstructural features through the formation of rosettes on the surface (Figure 9).

Silicon Carbide (SiC)

Silicon carbide decomposes in an environment of pure hydrogen [32,33]. Reaction with hydrogen gas produces CH_4, SiH and SiH_4 as the predominant gaseous species with the stable solid phase being silicon. The decomposition of silicon carbide in hydrogen is in accordance with:

$$SiC + 2H_2 = Si + CH_4 \text{ (g)}$$

In an environment of pure hydrogen the silicon carbide would decompose at an increasing rate with increasing total system pressure [31]. On exposure to intermediate levels of moisture, active attack of the SiC grains was observed. However, on exposure to high moisture levels a protective SiO_2 scale forms on the surface of the silicon carbide [31,34,35]. The scale provides protection against further oxidation, the effectiveness of which depends on the morphology and stability of the scale itself (Figure 10). The material experienced no degradation in room temperature flexural strength on exposure to wet hydrogen (saturated with moisture at room temperature) for 100 hours at temperatures of 1400°C (Figure 11). However, Hullum and Herbell [32] observed the strength of the material to degrade on prolonged exposure to an environment of dry hydrogen. The degradation resulting from an absence of protective oxide scaling in the dry hydrogen environment.

Silicon Nitride (Si_3N_4)

The reaction of silicon nitride with pure hydrogen results in the formation of pure silicon along with the generation of nitrogen, ammonia, SiH and SiH_4 as the major gaseous products. At elevated temperatures, above 1200°C the partial pressure of nitrogen is an order of magnitude greater than those of SiH and SiH_4, and hence is the predominant gaseous product. Furthermore, at elevated temperatures, the silicon nitride decomposes to silica and nitrogen, in accordance with the reaction [31]:

$$Si_3N_4 = 3 Si \text{ (s)} + 2 N_2\text{(g)}$$

Consequently, pure hydrogen will exert little influence on the degradation of silicon nitride at high temperatures. At temperatures below 1100°C, degradation of this material occurs due to reaction with hydrogen resulting in the formation of SiH_4 and the release of gaseous nitrogen [31]:

$$Si_3N_4 + 6 H_2 = 3 SiH_4 + 2 N_2\text{(g)}$$

156

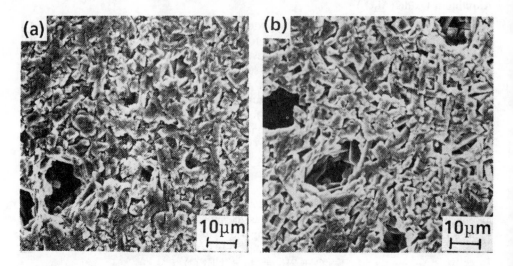

Figure 8.   Micrographs showing surface morphology of mullite (Ref. 31):
(a) as-received, and (b) exposed 100 hours to wet H_2 at 1100°C.

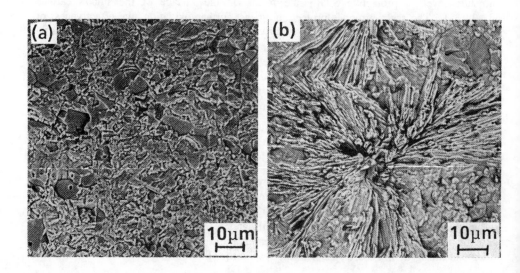

Figure 9.   Micrographs showing the morphology of aluminum oxide (Al_2O_3)
(Ref. 31)
(a) as-received, and (b) exposed 100 hours to wet H_2 at 1400°C.

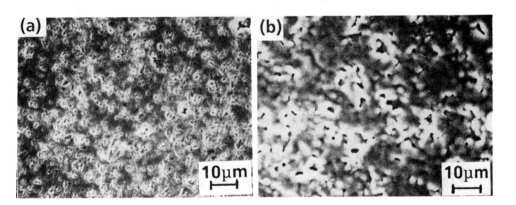

Figure 10. Optical micrographs showing surface morphology of SiC exposed
to hydrogen at 1300°C, showing (Ref. 31)
(a) as-received microstructure, and (b) protective SiO_2 film

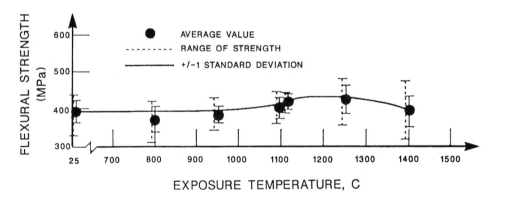

Figure 11. Effect of elevated temperature exposure to hydrogen, for 100
hours on room temperature strength of silicon carbide (Ref. 31).

In the presence of moisture the stability of Si_3N_4 is characterized by three distinct stability regimes [31], each region having its own thermodynamically stable solid phase. The flexural strength of silicon nitride following exposure to wet hydrogen for 100 hours at temperatures up to 1400°C, degrades only after exposure to temperature greater than 1200°C (Figure 12). The loss in strength also occurs when exposed to non-hydrogen containing environments and was attributed to degradation of the protective SiO_2 scale on the surface of the silicon nitride.

Silicon Nitride Joints with Metallic Layers

Silicon nitride in combination with metallic parts is contemplated for use in automobile engines. It is necessary to understand the influence of environment on the silicon nitride/metal system since the lifetime of the joint may be reduced and degradation of mechanical properties can occur. Combination of silicon nitride and aluminum as constituents of a joint produces a strong interface for structural purposes [36]. Suganuma and co-workers [36] evaluated environmental effects on mechanical properties of silicon nitride joints with metallic interlayers and having strong interfaces. In comparison with strength in argon, an inert environment, the strength of pure silicon nitride degraded in water. However, the degradation in strength of the silicon nitride joint was more severe than that of the pure ceramic (Figure 13). The degradation in strength was attributed to differences in fracture paths of joints tested in the two environments. In the argon environment the fracture occurred in the Si_3N_4 away from the joining interface while fracture was almost always along the interface when tested in water. The difference in fracture behavior was rationalized by the phenomenon of stress corrosion. The presence of water degrades the strength of the interface by stress corrosion resulting in fracture initiation and propagation along the interface where defects and local concentrations exist [Figure 14]. There exists two competing mechanisms for influence of stress corrosion on ceramic/metal joints:

   (a)   The first mechanism is a direct effect on the interatomic bonding between a ceramic and metal, and

   (b)   the second mechanism, is concurrent degradation of the interfacial region along with the ceramic, metal and reaction layers.

Since failure almost always occurred at the ceramic (Si_3N_4) and since ceramics are highly sensitive to stress corrosion, ceramic/metal joints become equally sensitive. Therefore, control of the ceramic/metal interface is desired. This is achieved by careful control of processing conditions and also through the use of raw materials (ceramics) that are pure and contain minimum impurities.

SUMMARY and CONCLUDING REMARKS

An examination of the literature reveals that novel monolithic ceramics and ceramic matrix composites are attractive candidates as structural materials for components requiring high temperature exposure and environmental compatibility. Monolithic mullite is unstable and severely attacked by an environment containing hydrogen and water vapor at temperatures as low as 1100°C. The room temperature fracture strength of mullite degrades on prolonged exposure to hydrogen. The degradation of mullite in an hydrogen containing environment, at high temperatures, is attributed to preferential attack of the glassy grain boundaries. Alumina (Al_2O_3) loses only a fraction of its room temperature flexural strength

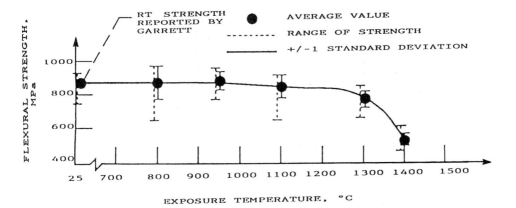

Figure 12. Effect of exposure to hydrogen at elevated temperatures, for
100 hours, on room temperature strength of silicon nitride
(Ref. 31).

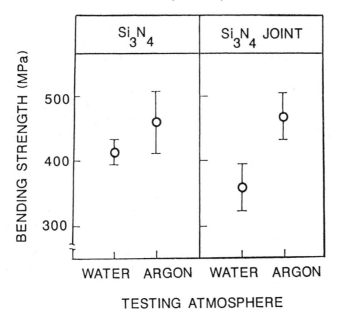

Figure 13. Effect of environment on bending strength of silicon nitride
(Ref. 36)

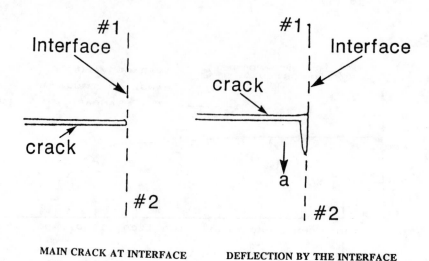

MAIN CRACK AT INTERFACE   DEFLECTION BY THE INTERFACE

Figure 14.   Schematic showing crack deflection at an interface:
(a) main crack at an interface, (b) deflection at interface.

on exposure to an environment of wet hydrogen.   The degradation is related to intrinsic changes in microstructural morphology.   Silicon carbide and silicon nitride decompose to form the silica phase on exposure to hydrogen at elevated temperatures.   The stable silica layer forms on the surface and imparts stability.   Degradation of the protective silica-based scale on the surface resulted in a loss of strength of the material.   Ceramic-metal joints degrade by stress corrosion in moist hydrogen-containing environments.   The degradation resulting from the conjoint action of weakening of the interface and preferential crack growth along the interface.   Environmental resistance of ceramic-metal interfaces is of concern in developing advanced composite systems.   Future research and development efforts on ceramic-metal interfaces need focus on:

(1)   the need to join using processes that produce minimum heat at the interface,
(2)   the need to join or braze metal-ceramic materials that have minimum diffusion along grain boundaries when exposed to high temperatures and aggressive environments, and
(3)   understand the behavior and segregation aspects of alloying elements when exposed to high temperatures and aggressive environments and their influence on mechanical behavior.

ACKNOWLEDGEMENTS

The authors extend their thanks and appreciation to Mrs. Sandra Collins for her timely and invaluable assistance in the preparation of this manuscript.

REFERENCES

1. Petrasek, D.W., McDaniels, D.L., Westfall, L.J. and Stephens, J.R.: **Metal Progress**, 1986, Vol. 130 (2), p. 27.

2. Petrasek, D. W. and Signorelli, R.A.: NASA Technical Memorandum 82850, 1981, NASA Lewis Research Center.

3. Petrasek, D.W., Signorelli, R.A., Caulfield, T. and Tien, J.K.: in Superalloys, Composites and Ceramics (editors: J.K. Tien and T. Caulfield), Academic Press, New York.

4. Stephens, J.R. and Nathal, M.V.: "Status and Prognosis for Alternative Engine Materials," NASA Technical Memorandum 100903, presented at the 6th International Symposium on Superalloys, TMS: Minerals, Metals, Materials Society, Warrendale, September 1988.

5. Das, G. and Vahldiek, F.W.: in Corrosion and Particle Erosion at High Temperatures (editors: V. Srinivasan and K. Vedula), Warrendale, PA, 1989, p. 531.

6. Das, G. and Vahldiek, F.W.: in Interfaces in Metal-Ceramic Composites I: Thermodynamics and Kinetics, (editors: R.Y. Lin and G.P. Martins), TMS: Minerals, Metals and Materials Society, Warrendale, PA, 1990.

7. Evans, A.G. and Faber, K.T.: **Journal of American Ceramic Society**, Vol. 67(4), 1984, p. 255.

8. Faber, K.T. and Evans, A.G.: **Acta Metallurgica**, Vol. 31(4), 1983, p. 565.

9. Marshall, D.B., Cox, B.N. and Evans, A.G.: **Acta Metallurgica**, Vol. Vol. 33(11), 1985, p. 2013.

10. Evans, A.G. and McMeeking, R.M.: **Acta Metallurgica**, Vol. 34(12), 1986, p. 2435.

11. Schoenlein, L.H., Jones, R.H., Henager, C.H., Schilling, C.H. and Gac, F.: Materials Research Society Symposium Proceedings, Vol. 120, 1988, p. 313.

12. Burger, W.G. and Weigel, C.W.: **IBM Journal Research and Development**, Vol. 27, 1983, p. 11.

13. Blodgett, A.J., Jr.: **Scientific American**, Vol. 249, 1986, p. 86.

14. Blodgett, A.J. and Barbour, D.R.: **IBM Journal of Research and Development**, Vol. 26, 1982, p. 30.

15. Evans, A.G., Ruhle, M., Dalgleish, B.J. and Charalambides, P.G.: in Metal Ceramic Composites (edited by M. Ruhle, A.G. Evans, M.F. Ashby and J.P. Hirth), Pergamon Press, Oxford, 1990, p. 345.

16. Sigl, L.S., Mataga, P.A., Dalgleish, B.J., McMeeking, R.M., and Evans, A.G.: **Acta Metallurgica,** Vol. 36, 1988, p. 945.

17. Hu, M.S. and Evans, A.G.: **Acta Metallurgica,** Vol. 37, 1989, p. 917.

18. Ashby, M.F., Blunt, F.J. and Bannister, M.: **Acta Metallurgica,** Vol. 37, 1989, p. 1847.

19. Evans, A.G. and Marshall, D.B.: Materials Research Society Symposium Proceedings, Vol. 120, 1988, p. 213.

20. Mittal, K.L. (editor) Adhesion Measurement of Thin Films, Thick Films and Bulk Coatings, ASTM STP 640, Philadelphia, 1978.

21. Derjaguim, B.V.: Recent Advances in Adhesion, Gordon and Breach, New York, 1971, p. 513.

22. Pauling, L.: General Chemistry, Third Edition, Freeman Publishers, San Francisco, 1970, p. 118.

23. Hondros, E.D.: in Physicochemical Measurements in Metal Research; Techniques for Metal Research Series (editor: R.A. Rapp), Interscience Publishers, 1970, New York, Vol. IV(2), p. 293.

24. Shieu, F.S., Raj, R. and Sass, S.L.: **Acta Metallurgica Materia,** Vol. 38 (11), 1990, p. 2215.

25. Agrawal, D.C. and Raj, R.: **Acta Metallurgica,** Vol. 37, 1989, p. 1265.

26. Rice, J.R. and Sih, G.C.: **Journal of Applied Mechanics,** Vol. 32, 1965, p. 418.

27. Dundurs, J.: **Journal of Applied Mechanics,** Vol. 36, 1969, p. 650.

28. Herbell, T.P., Hull, D. and Hallum, G.W.: in Hydrogen Effects on Material Behavior (editors: N.R. Moody and A.W. Thompson), TMS, Warrendale, PA, 1990, p. 351.

29. Crowley, M.S.: **Bulletin of American Ceramic Society,** Vol. 46, 1967, p. 679.

30. Chen, C.I.: The Reduction of Silica and Mullite in Hydrogen, PhD Thesis, Ohio State University, 1979.

31. Herbell, T.P., Eckel, A.J., Hull, D.R. and Misra, A.K.: in Environment Effects on Advanced Materials (editors: R.H. Jones and R.E. Ricker), TMS: Minerals, Metals and Materials Society, 1991, p. 159.

32. Hallum, G.W. and Herbell, T.P.: **Advanced Ceramic Materials**, Vol. 3(2), 1988, p. 171.

33. Fischman, G.S., Brown, S.D. and Zangvil, A.: **Materials Science and Engineering**, Vol. 72, 1985, p. 295.

34. Costello, J.A. and Tressler, R.E.: **Journal of American Ceramic Society**, Vol. 64(6), 1981, p. 327.

35. Gulbransen, E.A. and Jansson, S.A.: **Oxidation Metals**, Vol. 4(3), 1972, p. 181.

36. Suganuma, K., Niihara, K.. Fujita, T. and Okamoto, T.: in Metal- Ceramic Interfaces (editors: M. Ruhle, A.G. Evans, M.F. Ashby and J.P. Hirth), Pergamon Press, New York, 1989, p. 335.

Session IV

Structural Ceramics II

GRAIN-BOUNDARY STRUCTURE AND HIGH-TEMPERATURE STRENGTH OF NON-OXIDE CERAMICS

S. Tsurekawa and H. Yoshinaga

Department of Materials Science and Technology, Graduate
School of Engineering Sciences, Kyushu University, Kasuga Fukuoka, 816, Japan.

ABSTRACT

Grain boundary structures in pressureless sintered SiC with B+C sintering aids, hot isostatic pressed SiC without any sintering aids, and TiC were observed by means of high-resolution electron microscopy. Compression tests were conducted to clarify the effect of crystal structure on the high-temperature deformation behaviour of SiC. Main results obtained are as follows: (1) An amorphous-like layer about 0.5 nm thick was observed at grain boundaries in SiC. However, when the boundary was parallel to a low index plane and apparently of low energy, such a layer was not observed. These results strongly suggest that the amorphous-like layer is not an impurity layer but a relaxed structure. (2) A grain boundary layer such as observed in SiC was not observed in TiC, where grain boundary dislocations were periodically observed along grain boundaries that can be described by the DSC dislocation model. (3) There was no marked difference in high-temperature deformation behaviour between β-SiC, which satisfies von Mises criterion, and α-SiC, which does not satisfy the criterion. The similarity between the behaviour of β-SiC to that of α-SiC may arise from restricted dislocation motion due to dense stacking faults and twins produced in β-SiC.

INTRODUCTION

Grain boundary properties depend on its character, in particular the orientation relation between two adjacent crystals and the orientation of the grain boundary plane. Mechanical properties, for example, grain boundary sliding at high-temperature, are markedly affected by the grain boundary character. The recent development of high resolution electron microscopy has made it possible to associate the effect of grain boundaries on the mechanical properties with the grain boundary structures. It is well known that silicon carbide and silicon nitride, which are expected to be useful as structural materials at high-temperature, are difficult to densify by sintering without the addition of sintering aids.

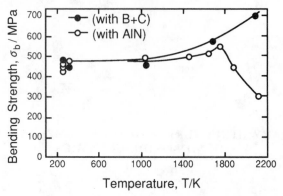

Figure 1. Temperature dependence of the bending strength for two types of sintered SiC[1].

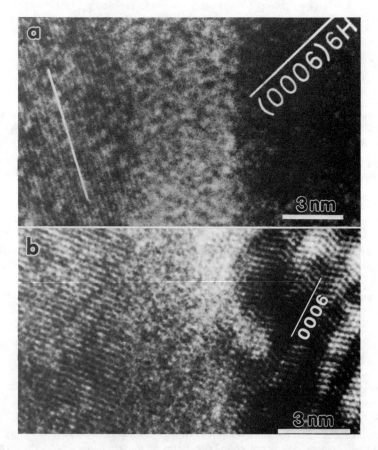

Figure 2. HRTEM photographs of a grain boundary in sintered SiC with B+C (a) and with AlN (b) [1].

However, sintering aids have a tendency to segregate at grain boundaries, resulting in a decrease of high-temperature strength. As shown in Fig.1, however, when boron and carbon are simultaneously added as sintering aids, the fracture strength of the sintered SiC does not decrease but rather increases at high temperatures[1]. As shown in Fig.2, high-resolution electron microscopic observation reveals the presence of an amorphous-like layer along the grain boundaries[1]. This layer is similar in appearance to that observed in the sintered SiC with AlN. Although the appearance is the same, the layer observed in SiC with B+C must be different from that in SiC with AlN, because the strength of the latter material does decrease at high temperatures. From this difference in the high-temperature strength, the layer observed in the former material is considered to be a relaxed structure formed to decrease the boundary energy[1]. However, to date there have been no convincing studies on the boundary structure in non-oxide ceramics. The objective of this paper is to reveal the grain boundary structures in silicon carbide with poor sinterability and titanium carbide with good sinterability by using high-resolution electron microscopy, and to clarify the mechanical properties, in particular the effect of crystal structure on the high-temperature deformation behaviour of SiC.

EXPERIMENTAL PROCEDURES

Specimens
The materials used here were pressureless sintered α-SiC (6H polytype: 96.4%), β-SiC (3C polytype: 98.1%) with additions of 0.37 mass% B and 3.1 mass% C, hot isostatic pressed β-SiC without any sintering aids, and pressureless sintered TiC without any additives.

High-resolution electron microscope observation
The high-resolution electron microscopic observation (HRTEM) of grain boundary structures was performed with a JEOL JEM-2000EX/T transmission electron microscope equipped with a top-entry double-tilt goniometer stage. The observation was conducted at 200 kV accelerating voltage. The spherical aberration coefficient Cs was 1.23 mm and the chromatic aberration coefficient Cc was 1.40 mm. These coefficients lead to resolving power of 0.25 nm at a Scherzer defocus of -66.7 nm; this is called the Scherzer resolution limit. An objective aperture of 6.5 nm^{-1} was used, which allowed beams up to 220 to pass through. Direct magnification was 6 $\times$ 10^5 and observations were conducted with a television monitor which magnifies the image 16 fold. Thin foil specimens of SiC and TiC were prepared by ion-thinning and twin-jet electropolishing techniques, respectively.

Compression test
Compression tests were carried out under a vacuum of approximately 1.3 mPa at temperatures from 1970 to 2320K and a strain rate of $3 \times 10^{-4} s^{-1}$ using an electrically controlled hydraulic machine based on the Shimadzu Servo Pulser EHF-2 type. For compression rods, sintered TiC was used. At the tips of the rods, melted TiC-10 mol% Zr, which is extremely strong at high-temperatures[2], was joined. The contact faces of the rods and specimens were painted with high purity graphite for lubrication and seizure resistance. The specimen was heated using radio frequency induction. The details of the heating method and temperature measurement were the same as those described in a previous paper[3].

The size of the specimens was approximately 2mm $\times$ 2mm $\times$ 3mm. The parallelism of the top and bottom sides of the specimens was so good that no leakage of light was observed when the specimen was put in a micrometer.

RESULTS AND DISCUSSION

Grain boundary structure in silicon carbide

Figure 3 is a HRTEM micrograph of a grain boundary in pressureless sintered α-SiC with B+C [4]. An amorphous-like layer is seen along the grain boundary. However, the layer is not truely amorphous but appears to have some microstructure. The thickness of the layer is approximately 0.5 nm; this is quite thin as compared with that reported by Ikuhara et al.[1], ca. 4nm. Since such an amorphous-like layer is not observed in high purity SiC prepared by CVD technique[5], the layer observed here could be another phase composed mainly of sintering aids and impurity oxgen. However, because the high-temperature strength of this material does not decrease, it is difficult to consider the layer as another phase. Lane et al.[6] reported that no Auger peak for boron was observed on the fractured grain boundary surface even when a B_4C particle is present nearby. Therefore, the boundary layer should not be considered another phase but a relaxed structure, called an extended grain boundary by Ikuhara et al.[1][7].

On the other hand, a low-index boundary, $(0001)//(10\bar{1}0)$, shows no sign of boundary extension[4]. From these observations, it may be concluded that a boundary in SiC with B + C tends to extend and reduce the energy when the coherency is poor, although the low-index and probably low energy boundaries do not extend.

Another kind of extended boundary was also observed. This was as wide as 3 nm and

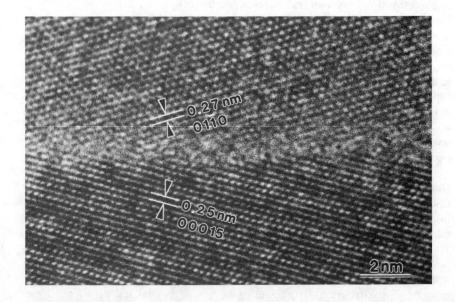

Figure 3. HRTEM photograph of a grain boundary in α-SiC with B+C[4].

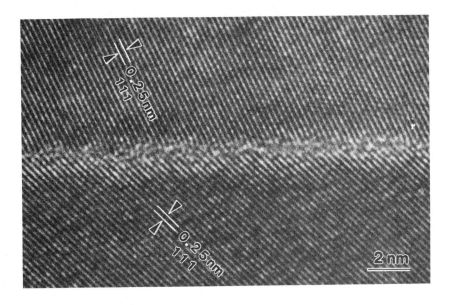

Figure 4. HRTEM micrograph of a grain boundary in hot isostatic pressed β-SiC without any additives[4].

composed of multiple twins with some lattice bending so as to achieve a fairly good coherency at the boundary[4].

A boundary structure observed in hot isostatic pressed β-SiC without any additives is shown in Fig.4 [4]. Again an amorphous-like layer is present that is similar in appearance to that observed in SiC with B + C. Although impurity oxgen can form this layer, the fact that the high-temperature strength of this material is very high[8] suggests that the layer should not be an impurity phase but should be a relaxed structure.

The boundary layer thickness observed in SiC by Ikuhara et al., ca. 4 nm, is much wider than those presented in this paper, about 0.5 nm. The question arises whether or not a structure relaxation can occur over a wide range of about 4 nm. A series of observations reveal that some strain exists around the boundary and could make the boundary extension look wider. The wide extension observed by Ikuhara et al. could originate from this strain effect. The second question is whether the amorphous-like structure observed in foils thinned by ion-milling has been formed by the milling or is a true intrinsic structure. In order to answer this question, a foil thinned by ion-milling was annealed at 1273K for 7.2 ks. No definite sign of a structure change was observed as a result of the annealing[4]. The activation energy for carbon and silicon self-diffusion in α-SiC has been reported to be 715 kJ/mol[9] and 697 kJ/mol[10], respectively. Assuming that the activation energy for boundary diffusion is 0.5 times as large as that for lattice diffusion, at the annealing temperature, it takes 47 s and 20 s for one jump of carbon and silicon, respectively, if the frequency term is assumed to be 10^{13}/s. Then, the jump number during the annealing time should be as large as 153 for carbon and 360 for silicon. Therefore, even if the layer were formed by ion-thinning, the jump numbers should be large enough for the atom rearrangement in the layer. However, no definite change in the boundary structure was observed. This could indicate that the amorphous-like structure is a true intrinsic one.

Grain boundary structure in TiC

Titanium carbide is a readily sinterable material, and can be densified without any additives. Figure 5 is a high-resolution image of the <110> tilt boundary with a misorientation angle of ϕ = 52° [11]. The boundary of $\phi = 52°$ is close to the Σ 11 coincidence boundary ($\phi =$ 50.48°), which is regarded to be a stable boundary in the fcc structure. In the figure, there is no definite sign of a boundary extension. The boundary structure is similar to that observed in metals such as Au [12] or Mo [13]. Moreover, some dislocations which are believed to produce the tilt angle are seen to lie on the boundary at close to an equi distance of about 4.5 nm. Thus, whether or not the DSC dislocations for Σ 11 can accomodate a small deviation of 1.2° from the exact coincidence orientation will be discussed below.

A schematic explanation of the DSC dislocation model for a Σ 11 coincidence boundary is shown in Fig.6[11]. From observation, it is found that the boundary plane is parallel to (001) on one of the adjacent crystals, and coincidence sites lie on the boundary. Thus, assuming that $\vec{b}_1$ dislocations ($\vec{b}_1 = (a/11)(\overline{3}1 1)$, a: lattice constant) produce the small deviation of the tilt angle $\Delta \phi$ from Σ 11, the value of $\Delta \phi$ estimated from the dislocation spacing is 1.5° . This agrees well with the measured value of 1.2° . Therefore, the boundary structure in TiC can be described by the DSC dislocation model. Further, the expected boundary layer was not observed even in incoherent boundaries in TiC[11]. From these observations, the grain boundary in TiC is thought not to differ greatly from that in metals, despite the highly covalent bonding.

Figure 5. High-resolution image of a grain boundary in TiC[11]. The arrows in the figure indicate boundary dislocations.

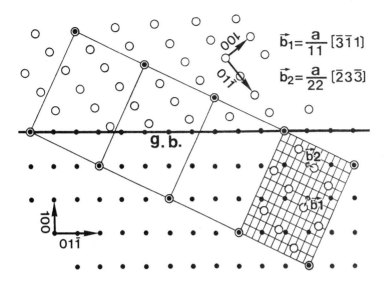

Figure 6. Coincidence sites and DSC lattice for Σ 11 <110> tilt boundary showing
Burgers vectors of DSC dislocations[11].

Effect of crystal structure on the high-temperature deformation behaviour of silicon carbide.

It is well known that many polytypes occur in SiC. However, they can be classified into two groups; one is β -SiC with a cubic structure and the other α -SiC with a hexagonal structure. In β -SiC, 12 {111}<1$\bar{1}$0> slip systems are activated at high-temperature. If five independent slip systems operate, polycrystals can be ductile. This is called in general von Mises criterion. At high temperatures where dislocation motion is thermally activated, β -SiC is expected to become ductile. On the other hand, the primary slip system in α -SiC is the (0001)<11$\bar{2}$0> basal slip. There are only two independent slip systems in this case. Thus, α -SiC is believed to have low ductility even at high temperatures.

Figure 7 shows stress-strain curves of α - and β -SiC sintered with B and C compressed at 2060K and an initial strain rate of 3×10^{-4} s^{-1}[14]. The materials used here have been sintered under the same conditions. Contrary to expectation, there is little difference in the stress-strain curves. At the peak stress, cracks were formed even in β -SiC, which could have caused the observed decrease in flow stress. Many stacking faults and microtwins were observed before and after the deformation[14]. The frequency of these defects increased after deformation. These defects could be the cause of the unexpected result. {111}<1$\bar{1}$0> slip in β -SiC should be restricted by these dense stacking faults and twins. It is believeaved, thererfore, that the ductility of β -SiC becomes low and β -SiC shows similar deformation behaviour to that of α -SiC. After deformation, X-ray diffraction analysis reveals that 6H and 4H polytypes have been produced in β -SiC[14]. TEM observations show that these hexagonal structures exist in the form of very thin twin-like layers[14]. Such twin formation and partial structure changes may be responsible for the absence of a significant difference between α - and β - SiC.

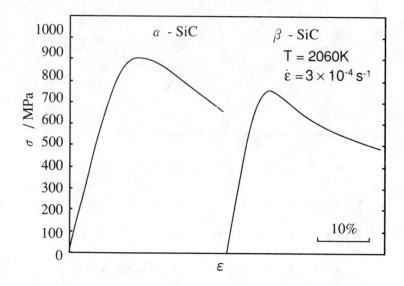

Figure 7. Stress-strain curves for α-SiC and β-SiC compressed at 2060K
and at a strain rate of $3 \times 10^{-4} s^{-1}$[14].

SUMMARY

In order to examine the concept of extended boundary proposed by Ikuhara et al., grain boundary structures in pressureless sintered SiC with B + C additives, hot-isostatic pressed SiC without any additives and pressureless sintered TiC were observed by means of high-resolution electron microscopy.   Compression tests were conducted in order to clarify the effect of crystal structure on the high-temperature deformation behaviour of SiC. The results obtained are as follows:

(1) In sintered SiC with B + C additives and without any additives, boundary layers as wide as approximately 0.5 nm were observed at high energy boundaries, although no extension was observed at low energy boundaries. These results support the concept of extended boundaries. However, the extensions were much smaller than those reported by Ikuhara et al..

(2) In TiC, no definite sign of boundary extension was observed. On the boundary of $1.2°$ off from $\Sigma 11$ coincidence relationship, boundary dislocations were observed and the deviation angle can be described by the DSC dislocation model.

(3) There was little difference in the high-temperature deformation behaviour between β-SiC, which satisfies the von Mises criterion, and α-SiC, which does not satisfy the criterion. The probable explanation is the dense stacking faults and microtwins produced in β-SiC.

ACKNOWLEDGEMENT

The authors would like to express their thanks to Dr. Y. Ikuhara, the Japan Fine Ceramics Center, for the supply of pressure-less SiC and to Dr. I. Tanaka, the Institute of Scientific and Industrial Research, Osaka University, for the supply of hot isostatic pressed SiC. The present work was financially supported by the Grant-in-Aid for Scientific Research, the Ministry of Education, Science and Culture, Japan. The support is also very much appreciated.

REFERENCE

1. Ikuhara, Y., Kurishita, H. and Yoshinaga, H., Grain boundary and high temperature strength of sintered SiC. Yogyo-kyokai-shi, 1987, 95, 638-645.
2. Tsurekawa, S., Nakashima, M., Murata, A. and Yoshinaga, H., Solid solution hardening of titanium carbide by niobium and zirconium at high-temperature. J. Japan Inst. Metals, 1991, 55, 390-397.
3. Kurishita, H., Nakajima K. and Yoshinaga, H., The high-temperature deformation mechanism in titanium carbide single crystals. Mater. Sci. Eng., 1982, 54, 177-190.
4. Tsurekawa, S., Nitta, S. and Yoshinaga, H., High-resolution electron microscopic observation of grain boundaries in silicon carbide. Will be submitted to Mater. Trans., JIM.
5. Ichinose, H., Inomata, Y. and Ishida, Y., HREM analysis of SiC grain boundary structure. Yogyo-kyokai-shi, 1986, 94, 415-418.
6. Lane, J. E., Carter, Jr., C. H. and Davis, R. F., Kinetics and mechanisms of high-temperature creep in silicon carbide: , sintered α -silicon carbide. J. Am. Ceram. Soc., 1988, 71, 281-95.
7. Ikuhara, Y., Kurishita, H. and Yoshinaga, H., Grain boundary structure and mechanical properties of covalent-bonded ceramics. Proc. 2nd Inter. Conf. on Interfaces in Polymer, Ceramic and Metal Matrix Composites, Elsevier Sci. Pub. Comp., 1988, 673-684.
8. Tsurekawa, S., Hasegawa, Y. and Yoshinaga, H., High-temperature deformation behaviour of hot-isostatic-pressed β -silicon carbide without any additives. Will be submitted to J. Japan Inst. Metals.
9. Hong, J. D. and Davis, R. F., Self-diffusion of carbon-14 in high-purity and N-doped alpha-SiC single crystals. J. Am. Ceram. Soc., 1980, 63, 546-52.
10. Hong, J. D., Davis, R. F. and Newbury, D. E., Self-diffusion of silicon-30 in Alpha-SiC single crystals. J. Mater. Sci., 1981, 16, 2485-94.
11. Tsurekawa, S. and Yoshinaga, H., Grain boundary structures in titanium carbide. Will be submitted to J. Japan Inst. Metals.
12. for example, Kvan, E. P. and Balluffi, R. W., Observations of hierarchical grain-boundary dislocation structures in [001] symmetric tilt boundaries in gold. Phil. Mag., 1987, 56, 137-148.
13. Tsurekawa, S., Tanaka, T. and Yoshinaga, H, Microstructure of <110> symmetric tilt-boundaries in molybdenum. To be published in J. Japan Inst. Metals.
14. Tsurekawa, S., Sato, K., Sakaguchi, Y. and Yoshinaga, H., Effect of crystal structure on high-temperature deformation behaviour of silicon carbide. Will be submitted to J. Japan Inst. Metals.

PROCESSING PHENOMENA FOR RECRYSTALLIZED SILICON CARBIDE

JOCHEN KRIEGESMANN
Fachhochschule Rheinland-Pfalz,
Abt. Koblenz, Fachbereich Keramik
Rheinstraße 56, D-5410 Höhr-Grenzhausen, Germany

ABSTRACT

The sintering behaviour of recrystallized silicon carbide and
the influence of grain size distribution, green density, SiO_2-
and C-content are described. Process engineering aspects are
derived from the evaporation/condensation sintering model
presented.

INTRODUCTION

Due to its porosity (approx. 20%), grain size (approx. 100
μm), shrinkage behaviour at firing (no shrinkage occurrs), and
applications (kiln construction), recrystallized silicon carbide (RSiC) is comparable to most refractory materials. With
respect both to its purity (> 99%) and to the bonding of the
crystals (direct grain/grain-joints), however, RSiC resembles
modern structural ceramic materials.

RSiC has a long hitory: the first patent was granted at
the turn of the century (1). But it is only in the last few
years that RSiC has been used commercially to a large extent,
probably because high capacity, high temperature sintering furniture has only been available relatively recently. A typical
application of RSiC is in the field of ceramic kiln construction where it is used as kiln furniture, for burner nozzles,
for kiln lining in aggregates for the firing of various ceramic
products, or as rollers in roller kilns.

The properties of RSiC differ from those of cordierite and
of silicon bonded and nitride bonded SiC in its higher refrac-
toriness, better thermal conductivity, and higher strength.
Therefore, RSiC parts with thinner walls can be produced, which
leads to more economical firing.

In comparison with silicon infiltrated silicon carbide
(SiSiC), RSiC has a higher thermal stability above 1250°C.
Below 1250°C, however, SiSiC is highly superior to RSiC, because it has - due to its higher density - a higher stability,
a higher thermal conductivity, and better thermal shock resistance. Therefore, RSiC should only be used at temperatures
above 1250°C. At lower temperatures, better and cheaper
materials are preferable.

However, for temperatures above 1250°C there is a type of
SiC that appears to be superior to RSiC with respect to
material properties, namely pressureless sintered silicon carbide (SSiC). For the time being, however, it is - for reasons
of costs and processing - not yet available for ceramic kiln
construction.

SINTERING BEHAVIOUR

The salient feature of RSiC is that no shrinkage occurs despite
obvious neck growth. Consequently, the density of the product
does not change on firing. Thus the material's strength is
relatively low. The advantage of shrinkage free sintering is

that large parts can be produced with high dimensional accuracy without distortion. Furthermore, the volume of this material does not change under working conditions.

The author has suggested an evaporation/condensation mechanism as a sintering model (2-5). It is not simple evaporation (or rather sublimation) of SiC, because SiC molecules are present only in a small amount in the gaseous phase. But SiC crystals decompose in the presence of SiO_2. Generally a thin layer of SiO_2 adheres to the SiC grains as a product of oxidation, which evaporates at higher temperatures as follows:

$$SiO_2(g) \longleftrightarrow SiO(g) + 1/2\ O_2(g)$$

i.e. the surface is exposed so that SiC can react with the SiO as follows:

$$SiC(s) + SiO(g) \longleftrightarrow 2\ Si(g) + CO(g)$$

The following reaction results from the combination of the above reactions:

$$SiC(s) + SiO_2(s) \longleftrightarrow 2\ Si(g) + CO(g) + 1/2\ O_2(g) \qquad \langle 1 \rangle$$

These reactions are also occurring during the synthesis of SiC powder via carbothermal reduction (6).

It is interesting to note that on the left side of $\langle 1 \rangle$ there are only solid products and on the right side only gaseous ones, i.e. the forward reaction (from left to right) can be considered an evaporation (sublimation), and the counter reaction (from the right to the left), a condensation (re-sublimation).

The driving force of the evaporation/condensation mechanism (7) is the vapour pressure as a function of the sur-face curvature. The surfaces of the particles are generally aligned in a convex way so that on the surface the vapour pres-sure is relatively high. The finer the particles, the more

convex the curvature and the higher the vapour pressure above
the particles. In the areas where the particles are in contact, the curvature is mainly concave. Here the vapour pressure is relatively low. A concave shape is formed when two
large particles come in contact with each other, which results
in particularly low vapour pressure. Between the areas of high
and low vapour pressure there is a steady transport of the gas
phase, in which the convex areas are sources of the gas
transport and the concave areas are sinks.

If a powder charge shows a bi-modal particle size distribution with a gap (8), i.e. if the powder consists of both
coarser and finer grain fractions, so that the finer particles
can occupy the space between the coarse particles, the fine
particles will decompose during sintering. The product of this
decomposition deposits as a sublimate at the points of contact
of the coarse particles. In this way new grain boundaries are
formed at the points of contact thereby strengthening the
microstructure. The sintering process is finished as soon as
the fine powder particles have completely disappeared leaving
behind pores. If the gas phase transport is limited locally
only to the sintered body, the green body and the sintered body
have the same density because the centres of the coarse particles have coalesced during the sintering process (7).

Figure 1 shows schematically both an idealized green body
with a bi-modal particle distribution and the corresponding
sintered body. It should be noted that there is no formation
of new crystals with the evaporation/condensation mechanism,
but only a growth of particles already present. While this indicates that the term "recrystallized SiC" is a misnomer, it
has gained general acceptance.

A bi-modal particle size distribution for a batch of RSiC
powder composition in practical experience, usually contains
about 70% of coarse grains ("skeleton forming grains") and
about 30% fine grained material ("bonding grains"). The coarse
grains consist of particles about 100 µm in size, with a grain

size distribution as narrow as possible, and the fine grains
have a wide particle size distribution from 10 μm down to the
submicron range.

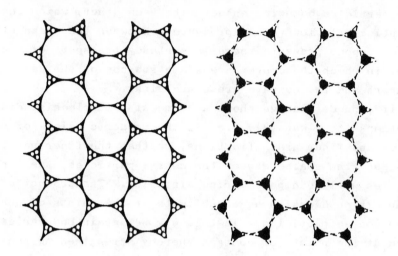

Figure 1. Schematic and idealized diagram of a green compact
(left) with a bi-modal particle size distribution and corresponding sintered body (right) under the conditions of an
evaporation/condensation mechanism (4-5).

By reducing the average grain size of the fine grain fraction, the sintering temperature can be further reduced. That
is, the soaking time can be shortened (9), due to the acceleration of the decomposition process by the finer SiO_2 grains, because in general the higher the convexity of the SiC-particles
and the higher their SiO_2 content the finer they are (5).
Figure 2 shows the microstructures of a slip-cast green body
and of the corresponding sintered body of a relatively fine-
grained RSiC (coarse grains about 20 μm).

Although the green body and the sintered body are very
similar with respect to their density, there are, however,
major differences in their microstructures. The grain boundaries that form during firing are mostly free of secondary

phases. This is because most of the impurities that occur in the raw material evaporate at high temperatures, since no sintering additives are used. The SiO₂-content in the sample also is reduced during firing due to the graphite in the sintering kiln (see below).

 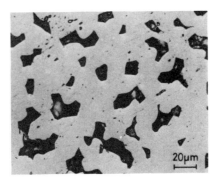

Figure 2. Optical micrograph of a green body with a bi-modal SiC particle distribution (left), and the corresponding sintered body (right); sintering temperature 2200°C in an argon atmosphere; sintering density: 79,1% of theoretical.

Because of the pure grain/grain bonding, the flexural strength does not drop, even at high temperatures (Figure 3), although the strength is relatively low due to the high porosity and the large particle sizes in comparison with SSiC. Also typical of RSiC is the transgranular fracture mode (Figure 4), even at low load velocities and high temperatures, which again is indicative of the quality of the RSiC grain boundaries (10-11).

The evaporation/condensation mechanism is not limited to a bi-modal SiC particle size distribution. Figure 5 shows the microstructure of a sintered body, for which a fine mono-modal SiC-powder with a specific surface of 10 m²/g was chosen as the raw material. No additives were used. This sample, too, sintered without shrinkage, which indicates a pure

182

evaporation/condensation mechanism. The microstructure of the
sintered body is characterized by idiomorphic crystals joined
by grain boundaries. Grain growth occurred during firing.

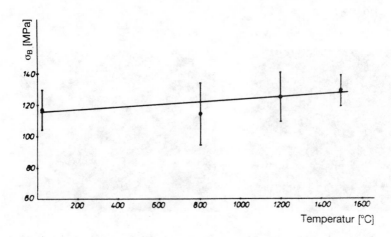

Figure 3. Temperature dependence of flexural strength for the
RSiC-sample in Figure 2.

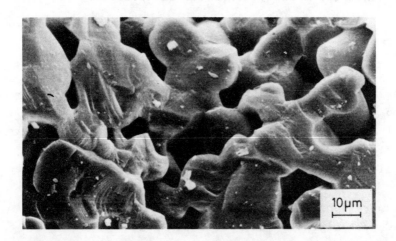

Figure 4. REM-micrograph of the fracture surface of the RSiC-
sample in Figure 2.

It should be noted that this fine powder sinters to 98% of theoretical density, if appropriate sintering additives are used (12).

Free carbon as a component of the SiC powder batch increases the formation of SiC in the presence of SiO_2 (2-5) according to the following equation:

$$SiO_2(s) + 3 C(s) \longleftrightarrow SiC(s) + 2 CO(g) \qquad \langle 2 \rangle$$

This equation is already familiar in connection with the "carbothermal reduction of SiO_2" for the mass production of SiC-powders (Acheson-process) (13). The influence of the evaporation/condensation mechanism should be reduced by the presence of free carbon in a RSiC powder batch, for it is no longer the decomposition but the formation of SiC that is favoured.

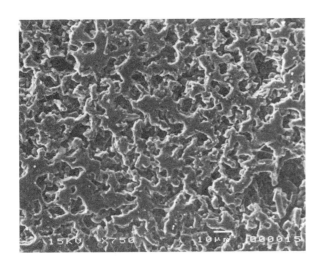

Figure 5. REM-micrograph of an isostatic pressed, sintered SiC-powder with a specific surface of 10 m^2/g; organic shape forming aids burnt out at 1000 °C in a H_2/N_2 gas mixture.; sintering temperature 2100 °C in an argon atmosphere; sintering density: 64,3 % of theoretical.

In order to check the validity of this statement, 7 mass% of carbon in the form of carbon black was added to the batch used as a sample in Fig.2. Fig.6 shows that the microstructures of the green body and the sintered body of the corresponding batch differ only slightly from one another. Some of the corners of the bodies become slightly rounded during the firing process, and thus some powder particles in the fine grain area can come into contact. The relation between the coarse and the fine grain parts has hardly changed, however. The evaporation/condensation mechanism was suppressed to a large extent. But as no further sintering mechanism, except for evaporation and condensation, is possible in the case of coarse grained powder, and if no additives that increased shrinkage are used, the microstructure cannot be strengthened during the firing process. Therefore the samples can be easily broken.

PROCESSING ASPECTS

Bi-modal grain size distribution defines the direction of mass transport during the sintering process. But the sintering test with fine mono-modal powder (Figure 5) has shown that bi-modal grain distribution is not absolutely obligatory for the sintering mechanism. Fine-grained SiC powder, however, seems to be necessary for the sintering mechanism, but it need only be present as a fraction of the powder batch. That means that bi-modal grain size distribution is only one factor for the composition of the raw material.

In order to obtain a product that is as homogeneous as possible, the green body should contain a large number of grains in contact with one another. For it must be assured that the gases formed by evaporation following the reaction <1> find enough points of contact for the condensation reaction during the firing stage. The type of point of contact, i.e. the degree of concavity that can be attained, appears to be of less importance. Optimization with respect to the numbers of contact points, however, does not seem to be possible only on the basis of a bi-modal grain size distribution (see below). Furthermore, a bi-modal grain size distribution cannot be arranged in practice by shaping as ideally as shown in Figure 1. Therefore not even a bi-modal grain distribution ensures uniformity in the points of contact of the skeleton grains. This statement contradicts the author's earlier opinion (2-5, 14) that, as with RSiC, the maintenance of a bi-modal grain distribution was obligatory for an optimized sintering process.

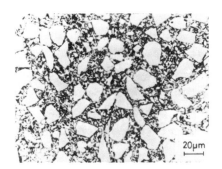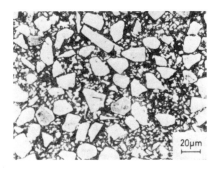

Figure 6. Optical micrograph of a green body with a bi-modal SiC particle size distribution containing 7 mass% carbon black (left) and of the corresponding sintered body (right); sintering temperature 2200°C in an argon atmosphere.

The requirement for a raw material batch, which is constructed in such a way that as many particles as possible come in contact with one another during shaping, is in agreement with the other requirement that the green density of RSiC should attain a maximum value. In the case of RSiC, it is of great importance that a high green density is attained because due to the evaporation/condensation mechanism, no further densification can occur during sintering. In general, a high degree of compression in the green state cannot be obtained using a continuous particle size distribution. Mixtures of fractionated particle sizes with miscibility gaps between the frac-

tions, as used for refractory bodies, behave more favourably.
Again bi-modal grain size distribution represents only one possible variation. The approaches suggested by Litzow (15) are
also feasible.

A high degree of densification combined with optimal sintering behaviour is very important to obtain strength as well
as oxidation resistance. It has been noted (9), however, that
strength increases with the decreasing size of skeleton forming
grains at constant sintering conditions and constant densification, whereas oxidation resistance improves with increasing
size of the skeleton forming grains. For applications, an appropriate compromise must be determined.

In practice, slip casting is used exclusively. This makes
possible the production of thin-walled, complexly shaped structural parts. What can be done to obtain high green densities
by using appropriately low viscosity slurries has already been
described (16) as well as what to do to facilitate removal
from the plaster mould. In Figure 7 the process to obtain RSiC
is shown schematically.

In comparison with other shaping processes, slip casting
makes it possible to eliminate organic additives in the RSiC-
batch. Consequently, after the drying process and before the
actual sintering, no condensation reaction or oxidation treatment is necessary.

If organic additives are not removed by evaporation or
oxidation treatment before firing, they are generally reduced
to carbon black during firing. Due to the graphite lining in
the kiln, the atmosphere is always a reducing one. Elemental
carbon hinders the sintering mechanism (according to equation
<2>) and hence reduces the stability of the product.

The uniaxially pressed RSiC-samples that were examined
first achieved only 25% of the strength of the slip-cast
samples (9, 14). The reason for the loss of strength in the
pressed samples was the share of free carbon that had been introduced through the forming aids. Less than 1% carbon clearly
reduces strength if the forming aids in the samples have not
been burned off before sintering.

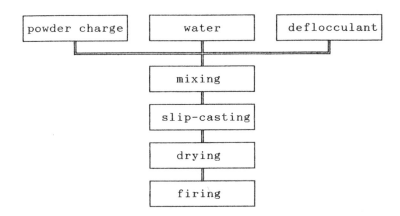

Figure 7. Processing of RSiC.

Basically, with respect to shaping RSiC, processes other than slip-casting can be used, e.g. axial pressing, isostatic pressing, or extruding. But these processes require pressing aids. If these organic additives evaporate at low temperatures or in reducing atmospheres such as camphor, dewaxing can be omitted. Most of the pressing aids, however, have to be carefully burned off in oxidizing atmospheres. This process does not seem to be very critical, because sintering is not impaired by the slight oxidation of SiC particles.

The firing of RSiC-particles can be carried out in resistance or induction furnaces. If possible, a static argon atmosphere should be used in order to prevent the gases that are produced through the evaporation reaction from being submitted to the condensation reaction. The above mentioned demand for a multitude of contact points in the green body insures that the gases condense within the sintered body and not anywhere else in the furnace. Graphite kiln furniture and linings can be used in the burning chamber as a getter for the Si and SiO gases formed superficially on SiC.

REFERENCES

1. Fitzgerald, F.A.J., US-Pat. 650 234 (priority 1899).

2. Kriegesmann, J., Keram. Z., 1986, **38**, 606-8.

3. Kriegesmann, J., powd. met. int., 1986, **18**, 341-4.

4. Kriegesmann, J., cfi/Ber. DKG, 1987, **64**, 301-3.

5. Kriegesmann, J., Interceram, 1988, **37**, 27-30.

6. Lee, J.G. and Cutler, I.B., Am. Ceram. Soc. Bull., 1975, **54**, 195-8.

7. Kuczynski, G.C., Trans. AME, 1949, **185**, 169-78.

8. Weaver, G.Q. and Logan, J.C., DE-Pat. 28 37 900 (priority 1978).

9. Kriegesmann, J. and Katheininger, A., 1980, unpublished work.

10. Kriegesmann, J., Keram. Z., 1989, **41**, 12-5.

11. Richter, H., Grellner, W. and Kriegesmann, J., Seminar: Festigkeit keramischer Werkstoffe, Stuttgart, 1980.

12. Märkert, J., Sperling, F. and Kriegesmann, J., 1989, unpublished work.

13. Ruff, O.O., Electrochem. Soc., 1935, **68**, 87-109.

14. Kriegesmann, J., Keram. Z., 1988, **40**, 943-7.

15. Litzow, R., Glastechn. Ber., 1930, **8**, 149-53.

16. Kriegesmann, J., Keram. Z., 1988, **40**, 857-63.

MICROSTRUCTURE AND ANELASTIC PROPERTIES OF Si_3N_4 CERAMICS

KAZUNORI SATO, KOHICHI TANAKA, AND YOSHINORI NAKANO
Department of Mechanical Engineering and Analysis Center
Nagaoka University of Technology
Nagaoka, Niigata 940-21, Japan

TSUTOMU MORI
Department of Materials Science and Engineering
Tokyo Institute of Technology
Nagatsuta, Midori-ku, Yokohama 227, Japan

ABSTRACT

The microstructure and mechanical properties of β-Si_3N_4 and α'/β-Si_3N_4 ceramics have been examined. Measurements of the Young's modulus at elevated temperatures by bend tests reveal that the anelastic properties of Si_3N_4 ceramics can be changed by reducing the amount of the glassy grain boundary phase. The addition of AlN as a sintering aid was found to strengthen the grain boundaries of dense compacts of Si_3N_4 due to the incorporation of elements of silicate glass, i.e. Si and O atoms, into β-Si_3N_4 grains. The observed decrease in the Young's modulus with increasing temperature was examined using a theory for polycrystalline anelasticity, which involves sliding grain boundaries against external stresses on the grain surfaces.

INTRODUCTION

Processing of Si_3N_4 ceramics to near-theoretical density requires the addition of sintering aids to promote liquid phase sintering. These additives react with the Si_3N_4 particles thereby forming a liquid phase during densification. However, it is usually accompanied by a certain amount of residual glassy phase at the grain boundaries. The thickness of this intergranular phase has been estimated to be 1-10 nm [1,2]. The presence of such a phase at the grain boundaries degrades the high temperature mechanical properties of Si_3N_4

ceramics.

A number of studies have been made to improve the high temperature strength of dense compacts of Si_3N_4 ceramics prepared by either hot pressing or hot isostatic pressing with Y_2O_3- and Al_2O_3-based additives [3-6]. However, the role of the intergranular phase on the deformation of Si_3N_4 ceramics at elevated temperatures has not been understood so far.

Therefore, we have examined the effect of the grain boundary phase on the deformation mode by bend tests up to 1400°C with both conventionally processed β-Si_3N_4 and hot pressed α'/β-Si_3N_4 composites. From these experiments, Young's moduli of these materials were measured at elevated temperatures in order to observe polycrystalline anelasticity with sliding grain boundaries.

EXPERIMENTAL PROCEDURE

Two types of dense compacts were prepared. One is a single phase β-Si_3N_4 obtained by hot isostatic pressing at 1750°C and 140 MPa in an argon atmosphere for 1 h. The other is two-phase composites of β-Si_3N_4 and Y-α'-sialon prepared by Toyota Central Research and Development Labs., Inc. [7]. No other crystalline phases were identified by X-ray diffraction analysis (XRD) for these compacts after densification. The compositions and phases of the specimens are given in Table 1.

TABLE 1
Chemical compositions of sintered Si_3N_4 ceramics

| Specimen | mol % | | | | Phases |
	Si_3N_4*	Y_2O_3	Al_2O_3	AlN	
A	94.0	3.8	2.1	–	β
B	91.0	0.9	–	8.1	β + α'(10 vol%)
C	77.0	2.3	–	20.7	β + α'(25 vol%)

* Ube, E-10, 1.04 wt% O

Phase identification and lattice parameter measurements were made with an X-ray diffractometer operated at 40 kV and 30 mA using monochromated Cu-Kα radiation on the polished surfaces of the specimens. All specimens were polished to a

surface roughness of 0.3 μm with diamond paste.   Care was
taken for the precise measurement of unit cell dimensions of
the a- and c-axes of β-Si$_3$N$_4$.   Profiles of (321), (212),
(312), (511), (322), (502), (203), and (303) lines were re-
corded at a scanning speed of 1/16 deg/min.   Peak positions
of these (n,k,l) Kα$_1$ lines were plotted against cos^{2}θ.
The diffractometer was calibrated against silicon powder,
a = 0.54308 nm.   A value of a = 0.54308 $\pm$ 0.00005 nm was
obtained.

Specimens used for the strength measurements were machined from the sintered bodies into rectangular bars approximately 4 mm wide, 3 mm thick, and 40 mm long.   All specimens
were polished with diamond paste on the tension side.   Bend
strength measurements were performed by a three-point bend
test with a 30 mm span, at a constant crosshead speed of
0.5 mm/min using a molybdenum fixture.   The tungsten mesh
furnace was raised to the selected temperature in 30 min and
held for at least 10 min prior to the bend test.   Total
dimensional variation during the bend test was measured with a
level gauge of 0.1 μm precision attached to the molybdenum
push rod.   The stiffness of the testing fixture was calibrated at each testing temperature by compressing a rigid Si$_3$N$_4$
block.   The Young's modulus, **E**, was determined from the
slope of the linear portion in the load versus the deflection
curve using the following equation:

$$E = \frac{l^3}{4bh^3}\left(\frac{P}{\delta}\right) \tag{1}$$

where **P** is the applied load; **l** is the span length; **b** is
the specimen width; **h** is the specimen thickness; and δ is
the deflection.

Fracture surfaces and microstructures of the specimens
were observed with a scanning electron microscope (SEM).
The microstructures were observed on polished sections after
etching in an 85% H$_3$PO$_4$ solution at 150°C for 15 min.

RESULTS AND DISCUSSION

The General Microstructure of the Si$_3$N$_4$ Ceramics
Figs. 1(a), (b), and (c) show the microstructures of specimens
A, B, and C, respectively.   These specimens exhibited hexago-
nal rod-like shapes of fine β-Si$_3$N$_4$ grains.   The average

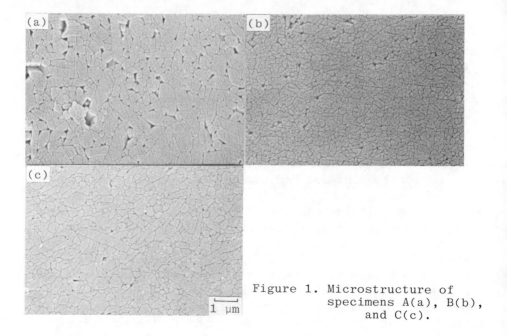

Figure 1. Microstructure of specimens A(a), B(b), and C(c).

grain size of these rod-like grains in specimens A and C appeared to be almost the same. However, the average grain size in specimen B was smaller than that in specimens A and C. Increasing the Y_2O_3 and Al_2O_3 contents in the starting mixture promote the grain growth of the rod-shaped β-Si_3N_4 via a solution-precipitation mechanism [8]. Thus, independent sintering tests were performed for mixtures with the same composition as specimens A and B by hot pressing at 1750°C. The result showed that the mixture corresponding to specimen A was fully densified but the one corresponding to specimen B was not. This reflects the difference in grain size between these specimens, which can be associated with the formation of a liquid phase during the densification process.

Lattice constants of β-Si_3N_4 in specimens A and C are summarized in Table 2. The lattice constants of a pure β-Si_3N_4 ceramic without sintering aids are also included for comparison. The values of the a- and c-axes in specimen A were almost the same as those in the pure β-Si_3N_4 ceramic[10] and smaller than those in specimen C. Al and O atoms are soluble in β-Si_3N_4 thus forming β'-sialon. Recent EDS data for β'-Si_3N_4 grains in a conventionally processed Si_3N_4 ceramic, prepared using 3.0 mol% Y_2O_3 and 7.5 mol% Al_2O_3, revealed that Al and O atoms, corresponding to an amount of

TABLE 2

Lattice constants of β-Si$_3$N$_4$ after densification

| Specimen | nm | |
	a-axis	c-axis
A	0.76048 ± 0.00002	0.29087 ± 0.00002
C	0.76136 ± 0.00005	0.29139 ± 0.00004
pure β-Si$_3$N$_4$**	0.76044	0.29075

** referred to [10]

about 6.8 mol% Al$_2$O$_3$, are incorporated in the β'-Si$_3$N$_4$ grains [6]. Jack has pointed out that β-Si$_3$N$_4$ reacts with equimolecular mixtures of Al$_2$O$_3$ and AlN during densification, so that the β' hexagonal unit-cell dimensions increase [9]. Thus, the same lattice constants in specimen A as those in the pure β-Si$_3$N$_4$ ceramic indicate that most of Al$_2$O$_3$ added in specimen A remains in the β-Si$_3$N$_4$ grain boundaries. This can be further supported by previous work in which Al tends to concentrate in intergranular glassy phases [1,6]. In contrast to this, the larger lattice constants of β-Si$_3$N$_4$ in specimen C than in the pure β-Si$_3$N$_4$ ceramic can be attributed to the formation of β'-sialon by the following reaction [7]:

$$(4-z)Si_3N_4 + zSiO_2 + 2zAlN \rightarrow 2Si_{6-z}Al_zO_zN_{8-z} \qquad (2)$$

This implies that the surface silica present on the starting Si$_3$N$_4$ powder reacts with the AlN additive, and a part of this silica is incorporated into the β'-Si$_3$N$_4$ grains. Thus the amount of the residual glassy grain boundary phase in specimens B and C is considered to be significantly reduced compared with that in specimen A.

Strength Evaluation

The Young's moduli for all three types of materials were measured as a function of temperature, as shown in Fig. 2. At room temperature these specimens showed values of 310-320 GPa, which correspond to those reported for hot isostatically pressed Si$_3$N$_4$ ceramics with Y$_2$O$_3$ [11]. The linear decrease in the Young's moduli for specimen A up to about 800°C and specimens B and C up to about 1200°C was in good agreement with that for hot-pressed Si$_3$N$_4$ ceramics [12]. However, the

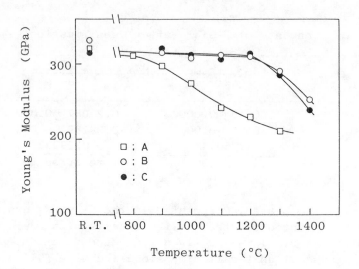

Figure 2. Young's moduli of materials as a function of temperature.

significant decreases in the Young's moduli at elevated temperatures for specimen A above 800°C and specimens B and C above 1200°C cannot be explained by the linear temperature dependency of the Young's modulus. This can be attributed to the anelastic deformation of polycrystalline Si_3N_4 ceramics with sliding grain boundaries at high temperatures, as explained by the Zener mechanism [13].

As stated in the previous section, the grain boundaries of specimens A, B, and C contain an amorphous phase that can soften at high temperatures resulting in viscous grain boundary sliding. The results of the lattice constant measurements of the β-Si_3N_4 grains imply that the amount of this glassy phase in specimens B and C was significantly reduced due to the incorporation into the β'-Si_3N_4 grains of the surface silica in the starting Si_3N_4 powder. Thus it can be inferred that specimen A showed poorer resistance to grain-boundary sliding with increasing temperature than specimens B and C. The decrease in the Young's moduli above 1200°C for specimens B and C was almost the same. This indicates that neither the average grain size nor the presence of the second phase, i.e. Y-α'-sialon, greatly affects the anelastic properties of Si_3N_4 ceramics.

The present results have shown that grain boundary sliding can play an important role in the evaluation of the elastic constants of sintered Si_3N_4 ceramics at high temperatures. Analytical forms of the effective elastic constants for a body

containing spherical sliding inclusions have been given by
Zener [13], Raj and Ashby [14], and Ghahremani [15]. Recently, Shibata et al. [16] have derived both the effective shear
modulus and Poisson's ratio of a body containing spherical
sliding inclusions and those of polycrystals with sliding
grain boundaries, using the average field theory [17,18].
According to the theory of Shibata et al., the effective shear
modulus, $\bar{\mu}$, for spherical grains is given as:

$$\bar{\mu} = \mu \frac{8(7 + 5\nu)}{161 - 65\nu} \tag{3}$$

where μ is the shear modulus without grain boundary sliding;
and ν is the Poisson ratio. Thus, the effective Young's
modulus, $\bar{E}$, for spherical grains is derived as:

$$\bar{E} = E \frac{5(21 + \nu) + 7(5\nu - 7)}{5(21 + \nu) + 7(3 + 5\nu - 10\nu^2)} \tag{4}$$

where E is the Young's modulus without grain boundary sliding. Similarly, $\bar{E}$ for cylindrical grains is given as:

$$\bar{E} = E \frac{12}{17 - 4\nu^2} \tag{5}$$

Using eqs. (4) and (5), these normalized Young's moduli,
$\bar{E}/E$, for spherical and cylindrical grains are calculated to
be 0.50 and 0.73 , respectively, when ν is taken as 0.25.
The measured value of Young's modulus at 1300°C for specimen A
was about 210 MPa, as shown in Fig. 2. Thus, the Young's
modulus normalized with the room temperature value becomes
about 0.65. This is in good agreement with the theoretical
prediction by Shibata et al.. Therefore, it can be concluded
that the shear traction acting on the surface of Si_3N_4 grains
due to sliding relaxes the external stresses, which is the
case with polycrystalline anelasticity with sliding grain
boundaries.
The bending strength of rupture in specimens A and C was
also measured as a function of temperature. The results,
shown in Fig. 3, further support the finding made in the
Young's modulus measurements that reducing the amount of an
amorphous grain boundary phase results in higher strength for

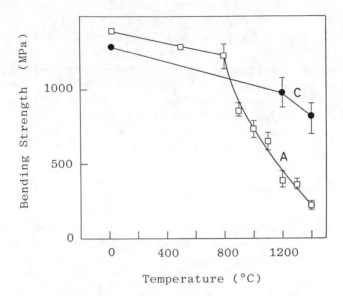

Figure 3. Bending strength of rupture as a function of temperature.

sintered Si_3N_4 ceramics.   This is probably caused by the enhancement of the cohesive strength of the grain boundaries over a wide range of temperatures.   The plot of the applied load vs the deflection at 1200°C for specimen A gave a nonlinear curve, indicating an occurrence of plastic deformation due to grain-boundary sliding, whereas that for specimen C did not.   This also indicates that improvement of mechanical bonding between the Si_3N_4 grains can be made by reducing the residual glassy phase at the grain boundaries.

Figure 4. SEM photographs of fracture surfaces in specimens A(a) and B(b) at 1400°C.

Figs. 4(a) and (b) show fracture surfaces close to the regions of fracture initiation at 1400°C in specimens A and C, respectively.   Specimen A clearly exhibited an intergranular fracture surface compared with specimen C.   In particular, the β-Si$_3$N$_4$ grains near the tensile side of specimen A showed distinct grain separation.   This is also evidence of plastic deformation by diffusionally accommodated grain boundary sliding at elevated temperatures.

CONCLUSIONS

Polycrystalline anelasticity due to grain-boundary sliding was observed for two types of Si$_3$N$_4$ ceramics.   This has been demonstrated by the measurement of the Young's modulus at elevated temperatures.   A decrease in the Young's moduli was observed for the β-Si$_3$N$_4$ ceramic sintered with Y$_2$O$_3$ and Al$_2$O$_3$ above 800°C and for the α'/β-Si$_3$N$_4$ composites sintered with Y$_2$O$_3$ and AlN above 1200°C.   This can be explained by a theory that deals with stress relaxation against external stresses on the surface of the grains.   In addition, it has been shown that the amount of the glassy grain boundary phase can be estimated from the precise measurement of the lattice con- stants of β-Si$_3$N$_4$.   The present results emphasize the signif- icant effect of grain boundary properties at elevated tempera- tures on the evaluation of both the bending strength of rup- ture and the Young's modulus, of Si$_3$N$_4$ ceramics.

REFERENCES

1. Ahn, C.H. and Thomas, G., Microstructure and grain bound- ary composition of hot-pressed silicon nitride with Yttria and alumina. J. Am. Ceram. Soc., 1983, **66**, pp. 14-17.

2. Clarke, D.R., High-temperature microstructure of a hot- pressed silicon nitride. ibid., 1989, **72**, pp. 1604-1609.

3. Tsuge, A., Nishida, K. and Komatsu, M., Effect of crystal- lizing the grain-boundary glass phase on the high-tempera- ture strength of hot-pressed Si$_3$N$_4$ containing Y$_2$O$_3$. ibid., 1975, **58**, pp. 323-326.

4. Loehman, R.E. and Rowcliffe, D.J., Sintering of Si$_3$N$_4$- Y$_2$O$_3$-Al$_2$O$_3$. ibid., 1980, **63**, pp. 144-148.

5. Govila, R.K., Strength characterization of yttria-doped sintered silicon nitride. J. Mater. Sci., 1985, **20**, pp. 4345-4353.

6. Cinibulk, M.K., Thomas, G. and Johnson, S.M., Grain-boundary-phase crystallization and strength of silicon nitride sintered with a YSiAlON glass. J. Am. Ceram. Soc., 1990, **73**, pp. 1606-1612.

7. Ukyo, Y. and Wada, S., Formation and stability of Y-α'-sialon co-existing with β'-sialon. In EURO-CERAMICS, ed. G. de With, R.A. Terpstra and R. Metselaar, Elsevier Applied Science Publishers, London, 1990, pp. 1566-1571.

8. Heinrich, J., Backer, E. and Bohmer. M., J. Am. Ceram. Soc., 1988, **71**, pp. C28-C31.

9. Jack, K.H., Sialons and related nitrogen ceramics. J. Mater. Sci., 1976, **11**, pp. 1135-1158.

10. Waring, J., Nat. Bur. Stand. (U.S.) Monogr., ASTM No.33-1160, 1981, **25**, Sec. 18.

11. Yeheskel, O., Gefen, Y. and Talianker, M., Hot pressing of Si_3N_4 with Y_2O_3. J. Mater. Sci., 1984, **19**, 745-752.

12. Gandhi, C. and Ashby, M.F., Fracture-mechanism maps for materials which cleave: f.c.c., b.c.c. and h.c.p. metals and ceramics. Acta. Metall., 1979, **27**, pp. 1565-1602.

13. Zener, C., Theory of elasticity of polycrystals with viscous grain boundaries. Phys. Rev., 1941, **60**, 906-908.

14. Raj, R. and Ashby, M.F., On grain boundary sliding and diffusional creep. Metall. Trans., 1971, **2**, pp. 1113-27.

15. Ghahremani, F., Effect on grain boundary sliding on anelasticity of polycrystals. Int. J. Solids and Structures, 1980, pp.825-845.

16. Shibata, S., Jasiuk, I., Mori, T. and Mura, T., Successive iteration method applied to composites containing sliding inclusions: effective modulus and anelasticity. Mech. Mater., 1990, **9**, 229-243.

17. Mori, T. and Tanaka, K., Average stress in matrix and average elastic energy of materials with misfitting inclusions. Acta. Metall., 1973, **21**, 571-574.

18. Mori, T. and Wakashima, K., Successive iteration method in the evaluation of average fields in elastically inhomogeneous materials. In Micromechanics and Inhomogeneity, the Toshio Mura Anniversary Volume, ed. G.J. Weng, M. Taya and H. Abe, Springer-Verlag, New York, 1990, pp. 269-282.

THE HIGH TEMPERATURE PROPERTIES OF Sc_2O_3 DOPED Si_3N_4

MASAKAZU WATANABE, TORU SHIMAMORI AND YOSHIRO NODA
Research & Development Center
NGK Spark Plug Co.,Ltd.
2808 Iwasaki, Komaki, Aichi 485, Japan

ABSTRACT

The relation between the high temperature properties of Si_3N_4 and sintering additives was investigated.The strength and oxidation resistance of hot-pressed Si_3N_4 fabricated from Si_3N_4 powder with two kinds of additives, Sc_2O_3 and another rare earth oxide, were evaluated. They were related to the ionic radius of the rare earth element added as the secondary sintering additive. As the ionic radius of the sintering additive with the exception of Sc_2O_3 became smaller,the ratio of the strength at 1400° C to the strength at room temperature (r.t.) and the oxidation resistance tended to become higher. Si_3N_4 doped with the oxides of Y, Dy, Er, and Yb as secondary sintering additives showed good high temperature properties.
Next, by using post-sintering technique, post-HIPed Si_3N_4 was investigated in terms of mechanical properties. It was found that the strength at r.t. and 1400° C of the sinter-HIPed Si_3N_4 doped with Sc_2O_3 and Y_2O_3 were the highest of all. The strength of the Sc_2O_3 and Y_2O_3 doped Si_3N_4 was about 800 MPa at r.t. and about 600 MPa at 1400° C.

INTRODUCTION

Si_3N_4 ceramics have been developed in recent years for structural applications at high temperatures.[1] Generally, Si_3N_4 ceramics can be fabricated by hot-pressing, normal sintering, gas pressure sintering, and hot isostatic pressing. These fabrication methods are mainly carried out for Si_3N_4 powder compacts with sintering additives. On the other hand, post sintering(PS), which starts with silicon powder, has been developed as one of the advanced sintering techniques for fabrication of Si_3N_4 ceramics. PS has several advantages over conventional sintering techniques. Among these, near-net-shape capability due to low shrinkage is the most well-known. Also, post-sintering methods which densify reaction-bonded Si_3N_4 (RBSN) are carried out by means of hot pressing[2], normal sintering, and HIPing [3],[4]. In the previous paper, we have reported that post-sintered Si_3N_4 doped with Sc_2O_3 showed good oxidation resistance.[6]

Based on the above results, in this paper the relation between high temperature properties and rare earth oxide additives of Si_3N_4 materials was investigated. In addition, the mechanical properties of Si_3N_4 materials post-sintered by HIPing were investigated.

EXPERIMENTAL PROCEDURE

Mixtures of commercial Si_3N_4 powder (B.E.T. about 7 m^2/g, α ratio > 95%, and oxygen content about 1 wt%) and 2.4 wt% Sc_2O_3+ 2.4 wt% of the second rare earth oxide were mixed in a solvent using a Si_3N_4 vessel and balls. As the second rare earth oxide, Y_2O_3, Pr_2O_3, Nd_2O_3, Sm_2O_3, Gd_2O_3, Dy_2O_3, Ho_2O_3, Er_2O_3, or Yb_2O_3 were used. These powder mixtures were hot-pressed under 20 MPa pressure at 1800°C for 2 hours. The hot-pressed samples were cut into specimens with dimensions of 3 by 4 by 40 mm using a diamond wheel. The strength at r.t. and 1400°C of the diamond-ground hot-pressed samples were measured according to JIS R-1601 and R-1604 (3-point bend test, span 30 mm). The weight gains of the hot-pressed samples were measured after oxidation in air and in a combustion gas (butane) at 1400°C for 10 hours. Element analysis and X-ray analysis of the oxidized samples were carried out.

Next, based on the above results, the mixtures of Si powder (B.E.T. about 3 m^2/g, purity ca. 99.8 %) + Sc_2O_3 and Si powder + (Sc_2O_3 + the other rare earth oxides) were mixed in a solvent using a Si_3N_4 vessel and balls. The oxides of Y, Dy, Er, and Yb were used as the other sintering additives. The amount of the additive Sc_2O_3 was 2.4 - 9.6 wt%, and the amount of total additives (Sc_2O_3 + other rare earth oxides) was 4.8 - 8.4 wt%. These mixtures were formed into square plates, which were then reaction bonded in a gaseous mixture of N_2 and H_2 at 1450°C. The reaction bonded samples were post-sintered. Post-sintering was carried out by sinter-HIPing at 200 MPa pressure in N_2 at 1800°C for 2 hours. The density of the post-HIPed samples was measured by Archimedes' method. The HIPed samples were cut into 3 by 4 by 40 mm specimens using a diamond wheel. The strength at r.t. and 1400°C of HIPed samples was evaluated by a 4-point bend test and the fracture toughness was evaluated according to JIS R-1604 (S.E.P.B. method). Oxidation resistance was evaluated in air at 1400°C for 100 hours. Creep resistance was also evaluated in air at 1400° C under a loading of 250 MPa by the 4-point bend method. Microstructures of etched samples were observed by S.E.M..

RESULTS AND DISCUSSION

INFLUENCE OF ADDITIVES ON THE HIGH TEMPERATURE PROPERTIES OF HOT-PRESSED Si_3N_4 FABRICATED FROM Si_3N_4 POWDER

STRENGTH

Fig.1 shows the 3-point bend strengths at r.t. and at 1400° C of hot-pressed samples doped with 2.4 wt% Sc_2O_3 plus 2.4 wt% of a second rare earth oxide. The flexural strength at r.t. of the sample doped with Sc_2O_3 + Pr_2O_3, of about 1000 MPa, was the highest of all the samples tested. The flexural strengths of the other hot-pressed samples at r.t. were about 800 MPa. The flexural strengths at 1400° C of all the hot pressed samples were about 500 MPa, and were almost independent of the sintering additives. The ratio of the flexural strength at 1400° C to that at r.t. appears to depend on both the type of the second rare earth oxide (other than Sc_2O_3) and the relation between the ratio of the flexural strength at 1400° C to that at r.t. and the ionic radius of the second rare earth oxide. This relation is plotted in Fig.2. As can be seen, the ratio of the flexural strength at 1400° C to that at r.t. tends to decrease with an increase in the ionic radius of the second rare earth oxide.

OXIDATION RESISTANCE

An oxidation test was carried out in air at 1400° C for 10 hours to evaluate the oxidation resistance of the hot-pressed samples. Also, in consideration of the possibility of their application as high temperature structural materials, the oxidation test was carried out in a combustion gas. Fig.3 shows the weight gain resulting from the oxidation test. There is a significant difference in the weight gain among the samples. The weight gains in the combustion gas were about the same as those in air. The data were rearranged in order of ionic radius of the rare earth elements as shown in Fig.4. It was found that the weight gain at 1400° C strongly depended on the ionic radius of the second rare earth element, and decreased linearly with decreasing ionic radius. However, the weight gains were almost the same for the compositions with ionic radii less than 0.9 Å. Fig.5 shows the distribution of the rare earth elements by electron probe microanalysis (EPMA) for cross sections of the oxidized samples doped with Sc_2O_3 + Pr_2O_3 and Sc_2O_3 + Yb_2O_3. In the Sc_2O_3 + Pr_2O_3 doped sample, which showed the highest weight gain, the elements Pr and Sc migrated to the oxidized layer where they segregated. From X-ray diffraction analysis of the oxide layer, it appears that silicates containing Pr or Pr + Sc exist in the oxidized layer. In the Sc_2O_3 + Yb_2O_3 doped sample, which showed the lowest

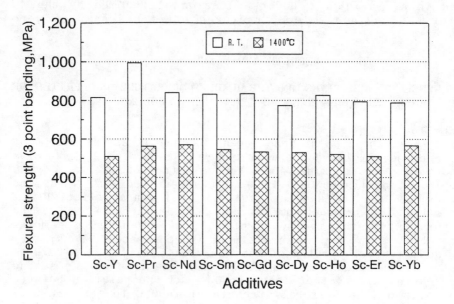

Fig.1 The flexural strength of hot-pressed Si_3N_4 at r.t. and at 1400° C.

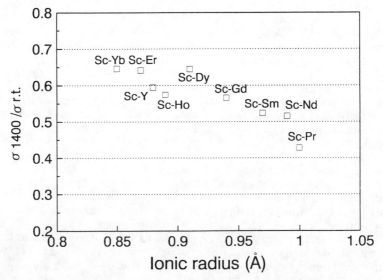

Fig.2 The ratio of the flexural strength at 1400° C to the flexural strength at r.t. of hot-pressed Si_3N_4 as a function of the ionic radius of the second rare earth element.

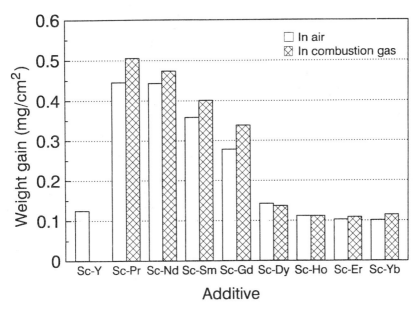

Fig.3 Weight gain of hot-pressed Si$_3$N$_4$ after oxidation in air
and in a combustion gas at 1400°C for 10 hours.

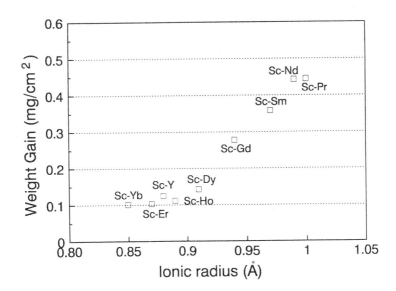

Fig.4 Weight gain of hot-pressed Si$_3$N$_4$ after oxidation in air at
1400°C for 10 hours as a function of the ionic radius of
the second rare earth element.

weight gain, neither Sc nor Yb migrated to any extent to the oxidized layer, and silicates containing Yb and Sc were not identified from X-ray diffraction analysis. Consequently, it is suggested that the rate limiting step for oxidation is the diffusion of the rare earth elements to the oxidized layer rather than oxygen diffusion in the oxidized layer. Next, the distribution of the sintering additives at the grain boundaries (triple point) of hot-pressed samples was analyzed by energy dispersive spectroscopy (EDS). The component of the sintering additives co-existed at the grain boundaries in all the samples. This result suggests that both Sc_2O_3 and the second rare earth oxide reacted with SiO_2 contained in the Si_3N_4 powder to form a complex oxide.

From these results, both the ratio of the high temperature strength to the room temperature strength and the oxidation resistance were found to depend on the ionic radius of the rare earth element in the sintering additive. It is believed that these high temperature properties depend on grain boundary properties, that is the properties of the grain boundary phase constituted from Sc_2O_3 - the second rare earth oxide - SiO_2 components. Furthermore, the grain boundary properties differ with different rare earth oxides. These differences affect the high temperature strength and oxidation resistance.

In general, the melting point of a compound is related to the lattice energy. The relation between the lattice energy and the ionic radius (distance between cation and anion)is shown below;[6]

$$U_L = - \frac{A \times Zi \times Zj \times e^2}{a} \times NA \left(1 - \frac{p}{a}\right)$$

where;
U_L is the lattice energy , A is the Madelung constant , Zi is the valence of the cation , Zj is the valence of the anion, e is the electric charge , a is the distance between the cation and anion , NA is Avogadoro's number, and p is the repulsive coefficient of the ion,

As shown in the above formula, the lattice energy increases with decreasing ionic radius of the elements in the compound. Since the melting point of a compound increases as the ionic radius decreases, it is believed that the degree of lattice energy of a complex oxide affects its high temperature properties.

INFLUENCE OF ADDITIVES ON THE HIGH TEMPERATURE PROPERTIES OF POST-HIPed Si_3N_4

FLEXURAL STRENGTH
Fig.6 shows the flexural strength at r.t. and at 1400°C of post-HIPed samples. The post-HIPed samples were nearly fully dense. The flexural strength at r.t. of the Sc_2O_3 + Y_2O_3 doped

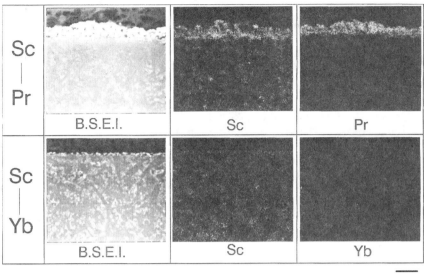

Fig.5 The distribution of rare earth elements in cross-sec-
tions of the samples after oxidation in air at 1400° C
for 10 hours.

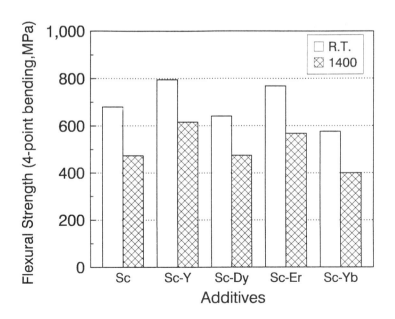

Fig.6 Flexural strength of post-HIPed Si₃N₄

sample, about 800 MPa, was the highest. Also, the flexural strength at 1400°C of this sample, 615 MPa, was higher than that of the other rare earth oxide doped samples. The strength at high temperature of the HIPed samples depended on the strength at room temperature and thus the flexural strength at 1400°C of the HIPed sample increased as the flexural strength at r.t. increased.

From observation of the fractured surface, the fast fracture mode of these samples showed brittleness. Fig.7 shows etched microstructures of the Sc_2O_3 doped sample and the Sc_2O_3 + Y_2O_3 doped sample, and the fracture toughness value of the two samples. The microstructure of the Sc_2O_3 + Y_2O_3 doped sample showed the most fibrous Si_3N_4 grains. The fracture toughnesses of the Sc_2O_3 + Y_2O_3 doped sample and the Sc_2O_3 doped sample were 6.1 $MPa \cdot m^{1/2}$ and 5.0 $MPa \cdot m^{1/2}$, respectively. The relation between the fracture toughness and the microstructure of the two samples can be seen in Fig.7.

OXIDATION RESISTANCE

The weight gain of the Sc_2O_3 doped sample after oxidation in air at 1400°C for 100 hours was the lowest. However, there was very little difference between the weight gains of all samples, which were 0.11 - 0.15 mg/cm^2.

CREEP RESISTANCE

Fig.8 shows the creep strain change of post-HIPed Si_3N_4 as a function of time under loading of 250 MPa at 1400°C. The creep strain of the Sc_2O_3 doped sample was smaller than that of the Sc_2O_3 + rare earth oxide doped samples. The four types of Sc_2O_3 + rare earth oxide doped samples showed almost equal creep resistance. The flexural strength at 1400°C of the Sc_2O_3 + Y_2O_3 doped sample was higher than that of the Sc_2O_3 doped sample. However, the creep resistance of the Sc_2O_3 + Y_2O_3 doped sample was poorer than that of the Sc_2O_3 doped sample, that is, a relation between the high temperature strength and the creep resistance was not found. It is considered that the difference between these two mechanical properties at high temperature is related to the properties of the grain boundary phases, that is, the viscoelastic behavior of the grain boundary at high temperatures. Therefore, it is necessary to investigate the difference between the two mechanical behaviors at high temperature from the standpoint of the viscoelastic behavior of the grain boundaries.

SUMMARY

The high temperature properties of Si_3N_4 doped with Sc_2O_3 and other rare earth oxides as sintering additives were investigated.

The experimental results are summarized below:

Sc$_2$O$_3$ Sc$_2$O$_3$ + Y$_2$O$_3$

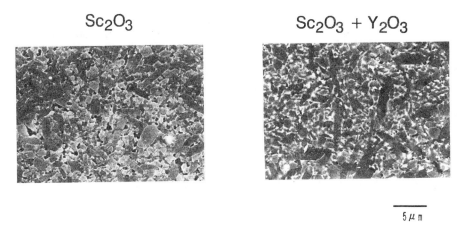

5 μm

Fig.7 Microstructure of post-HIPed Si$_3$N$_4$

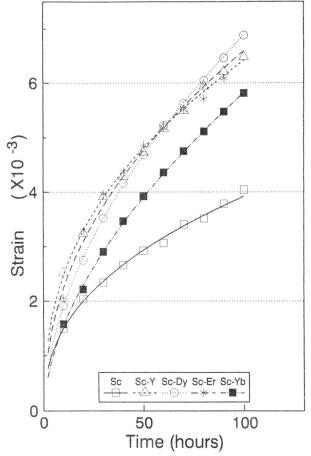

Fig.8 Creep behavior of post-HIPed Si$_3$N$_4$ doped with rare earth oxides under loading of 250 MPa at 1400° C.

1. The high temperature properties (the ratio of the strength at 1400°C to the strength at r.t. and the oxidation resistance) of Si_3N_4 fabricated by using Si_3N_4 powder as the starting material were strongly related to the ionic radius of the rare earth element. They improved with decreasing ionic radius of the rare earth element.
2. The flexural strength at r.t. and at 1400°C of post-sintered Si_3N_4 doped with Sc_2O_3 and Y_2O_3 after post-HIPing was the highest in this study. The flexural strength at 1400°C increased with increasing flexural strength at r.t..
3. In contrast to the high temperature flexural strength, the creep resistance of Sc_2O_3-doped Si_3N_4 was superior to that of Sc_2O_3 + Y_2O_3- or Sc_2O_3 + Er_2O_3-doped Si_3N_4.

ACKNOWLEDGEMENT

This work was performed under the management of Engineering Research Association for High Performance Ceramics as a part of the national R&D project " Research and Development of Basic Technology for Future Industries " supported by NEDO (New Energy and Industrial Technology Development Organization).

REFERENCES

1. Hattori, Y ., Tajima, Y., Yabuta, K., Matsuo, Y., Kawamura, M. and Watanabe, T., Gas Pressure Sintered Silicon Nitride Ceramics for Turbocharger Application. In Ceramics Materials and Components for Engines. ed. W. Bunk and H. Hausner , Deutche Keramische Gesellschaft , 1986, pp. 165-72.
2. Shimamori T., Kato T. and Matsuo Y., Development of oxidation resistance Si_3N_4 materials. In proceeding of the 1st symposium of the national R & D project, 1983, pp.47-65.
3. Ziegler, G and Wotting, G., Post-treatment of Pre-sintered Silicon Nitride by Hot Isostatic Pressing. Int. J. High Technology Ceramics, 1985,1,pp.31-58.
4. Heinrich, J., Backer E. and Bohmer M., Hot Isostatic Pressing of Si_3N_4 powder Compacts and Reaction-Bonded Si_3N_4. J. Am. Ceram. Soc., 1988,C-28-C-31.
5. Shimamori T., Development of Oxidation Resistant Silicon Nitride Ceramics by Post Sintered Technique., In proceeding of the 4th Symposium of the national R & D project, 1986, pp. 103-114.
6. Zalkin A. and Templeton D.H., J. Amer. Chem. Soc., 1953, 75, pp. 2453

THE LIFETIME OF SILICON NITRIDE LIMITED BY CAVITY NUCLEATION IN THE INTERGRANULAR GLASSY PHASE

JIANREN ZENG, ISAO TANAKA and KOICHI NIIHARA

The Institute of Scientific and Industrial Research, Osaka University,

Ibaraki, Osaka 567, Japan

ABSTRACT

An attempt has been made in this paper to predict the high-temperature delayed fracture strength of Si_3N_4 from the physical properties of its intergranular glassy phase. First, a new cavity nucleation model is introduced to evaluate the cavitation threshold of the glassy phase. A simple relation is then proposed to relate the cavitation threshold to the delayed fracture strength. The predictions agree with the delayed fracture strengths of Si_3N_4 containing different intergranular glassy phases. These results explain quantatatively why addition of SiO_2 does not cause any degradation in the delayed fracture strength of Si_3N_4 until a structure change occurs between 10 and 20 wt% SiO_2 additions, while the presence of trace impurities, such as fluorine, causes drastic degradation.

INTRODUCTION

Si_3N_4 is a promising material for gas-turbine components that are required to operate up to 1400°C under prolonged loads. However, the nucleation, growth, and coalescence of cavities in the intergranular glassy phases that lead to formation and/or propagation of cracks can cause Si_3N_4 components to fracture [1-6]. Because the cavitation threshold of the intergranular glassy phases represents the resistance of Si_3N_4 to delayed failure or failure initiation at ordinary operating temperatures, it is important to know the potential of this resistance, and the factors that determine it.

There has been relatively little theoretical investigation of cavity nucleation in the glassy phase [1,4,5,7]. The use of surface energy, a macroscopic quantity, on an almost atomistic scale in these models, is considered to be inappropriate when athermal defects are present. As an alternative, we propose that athermal defects are cavitation sites with locally reduced activation energy for cavity nucleation, and we use a statistical approach to evaluate the influence of impurities on the cavitation threshold of SiO_2. Taking into consideration stress concentrations at large inhomogeneities, we propose an approach for relating to relate the cavitation threshold of glassy phases to the delayed fracture strength. The delayed fracture strength of Si_3N_4 containing different amounts of SiO_2 are examined, and compared with that of Si_3N_4 with traces of fluorine. The influence of grain boundary structure on delayed fracture and creep deformation is also discussed.

CAVITY NUCLEATION INVOLVING ATHERMAL DEFECTS

At high temperatures, the intergranular glassy phase in ceramics behaves viscously, allowing grain boundaries to slide over each other when a tensile stress is applied to them. The sliding results in hydrostatic tension in the glassy phase at triple points, as is schematically shown in Fig. 1(a). Cavities can form by vacancy coalescence. The intergranular glassy pockets are generally of sufficient size to accommodate a critical nucleus with spherical morphology [4]. As indicated in Fig 1(b), the maximum in free energy for the growth of this spherical nucleus occurs at $r_c = 2\,\gamma_1/p$ [7,8],

$$\Delta G_c = \frac{16\pi\,\gamma_1^{\,3}}{3p^2}$$

(1)

where r_c is the critical nucleus size of the cavity, γ_1 is the vapor/viscous–phase surface energy, and p is the hydrostatic tension in the glass pocket.

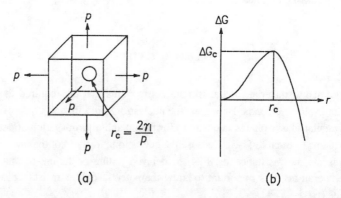

(a)   (b)

Figure 1 Schematic of (a) homogeneous cavitation in the glassy phase and (b) concomitant free energy changes.

The presence of surface energy, a macroscopic property, in the above equation automatically assumes structure continuity of the intergranular glassy phase. This assumption could become inappropriate when athermal defects are present. It is noted that in the case of SiO2, the activation energy for cavity nucleation is ~500 kJ/mol. While the presence of a single athermal defect, such as a ruptured Si-O bond, causes a local energy loss of 445 kJ/mol [9]. Clusters of ruptured Si-O bonds can further reduce the local energy.

In view of the structure discontinuity at athermal defects, they are considered here as nucleation sites associated with a reduced activation energy for critical cavity formation. To simplify this model, we assume that 1) the impurity is immobile and 2) the impurity content is too low for more than one defect to be involved in the formation of a critical nucleus. The difference between the formation energy of critical cavity nuclei calculated by using Eq.(1) and the formation energy at the defect i, ΔG_{ic}, is defined as ΔE_i. The value of ΔE_i, depends on the nature of the impurity and of the matrix glass, as well as on the defect size because a defect can consist of several several single defects. Thus;

$$\Delta G_{ic} = \Delta G_c - \Delta E_i \tag{2}$$

The number of the critical nuclei due to defect i, n_{ic}, and the total number of critical nuclei, n_c, in a glass pocket with a volume of V is given respectively by;

$$n_{ic} = V n_0 C_i \exp(\frac{-\Delta G_{ic}}{kT}) \tag{3}$$

and

$$n_c = \sum_{i=0}^{j} V n_0 C_i \exp(\frac{-\Delta G_{ic}}{kT}) \tag{4}$$

where V is the volume of the glass pocket and n_0 is the number of potential nucleation sites in a unit volume of glass. For a homogeneous nucleation process, the probability of developing a cavity should be the same for each molecule. Therefore, n_0 is assumed to be the number of molecules in a unit volume of glass, C_i is the concentration of defect i, j is the total number of different types of defects, T is the absolute temperature, and k is Boltzmann's constant.

The probability of adding a vacancy to a critical nucleus is given as [1,4,5]:

$$\omega(t) = \frac{8kT\gamma_1^2}{3\eta\Omega^{5/3}p^2} \tag{5}$$

where Ω is the molecular volume and η is the viscosity of glass. The cavity nucleation rate, $\dot{n}$, is determined by the product of $\omega(t)$ and the number of critical–size cavities [1,4],

$$\dot{n} = \frac{8kT\gamma_1^2}{3\eta\Omega^{5/3}p^2} \sum_{i=0}^{j} Vn_0 C_i \exp[-\frac{16\pi\gamma_1^3/3p^2 - \Delta E_i}{kT}] \qquad (6)$$

Note that when no athermal defect is present, $j = 0$ and $C_0 = 1$, Eq. (6) then reduces to:

$$\dot{n} = \frac{8kT\gamma_1^2}{3\eta\Omega^{5/3}p^2} Vn_0 \exp[\frac{-16\pi\gamma_1^3/3p^2}{kT}] \qquad (7)$$

PREDICTION OF DELAYED FRACTURE STRENGTH

It may be easier to clarify the role of cavity nucleation in the delayed fracture of Si_3N_4 than in other ceramics or metals. This is because the influence of preferential cavity nucleation sites such as slip bands [10,11], grain boundary ledges [12,13], and grain boundaries dislocations [14,15], all of which complicate cavity nuleation and thus the fracture process, is less dominant in Si_3N_4. According to Hsueh and Evans [16], stress is concentrated by a factor of ~2 at large inhomogeneities such as pores or oxidation pits. Since these inhomogeneities are difficult to avoid during processing or testing, when cavity nucleation is the controlling mechanism of delayed fracture, the delayed fracture strength of Si_3N_4, σ_{DFS}, can be estimated from the cavitation threshold of the corresponding intergranular glassy phase, p_{CT}, according to the following relation:

$$\sigma_{DFS} \approx 1/2\, p_{CT} \qquad (8)$$

The effect of stress concentration at triple points due to grain boundary sliding is ignored.

CALCULATION AND EXPERIMENTAL EVIDENCE

Effect of Glass Pocket Size

The cavitation rate is generally dominated by the exponential term in Eqs. (6) and (7), which gives a threshold–like behavior in response to the applied hydrostatic tension.

Figure 2 shows the threshold values for spherical cavities as a function of the size of the glass pocket, which is assumed to be cubic in shape. The cavitation threshold was calculated by assuming prescribed nucleation rates such that;

$$\dot{n} \cdot t = 10^{-1} \qquad (9)$$

Namely, one in ten glass pockets develop a cavity after a holding time of t (10^2, 10^5, and 10^8 s, respectively). The physical parameters pertinent to the present material system are given in Table 1. It is clear that the cavitation threshold of SiO_2 has little dependence on the volume of SiO_2 glass.

TABLE 1
Physical Properties of SiO2–based Glasses

Temperature (°C)	η (Pa·s) Pure	η (Pa·s) $C_l = 1\%$ $E_l = 445$kJ/mol	γ_l (J·m^2)	n_0 (SiO$_2$·m^{-3})	Ω (m^3·SiO$_2^{-1}$)
1000	5×10^{15}	1×10^{13}	0.3	10^{29}	10^{-29}
1100	2×10^{13}	2×10^{11}	0.3	10^{29}	10^{-29}
1200	2×10^{11}	2×10^{9}	0.3	10^{29}	10^{-29}
1300	2×10^{10}	1×10^{9}	0.3	10^{29}	10^{-29}
1400	5×10^{8}	1×10^{8}	0.3	10^{29}	10^{-29}
Ref.	17*	17**	18***		

* The viscosity of silica is assumed to be similar to glass containing 3 ppm OH⁻ ions.
** OH⁻ ions are assumed to completely rupture Si–O bond (445 kJ/mol).
*** The surface energy of silica is assumed to be unchanged when trace impurities are present.

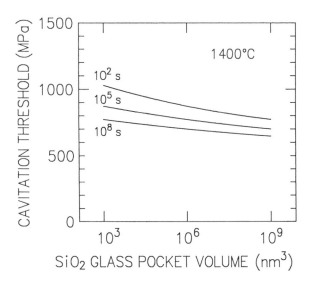

Figure 2 Calculated cavitation threshold for SiO_2 glass at 1400°C.

For comparison, Si_3N_4 containing different amounts of high-purity SiO_2 glass as the intergranular phase were used. These ceramics were fabricated by hot isostatic pressing. The details of fabrication, microstructures, and properties are reported elsewhere [19,20]. The delayed fracture strength and creep deformation of these materials at 1400°C were measured by a 4-point bending method, where stress was increased stepwise (50 MPa/12 h) until fracture occurred. The results are shown in Fig. 3. Note that the delayed fracture strengths of these ceramics are almost independent of their SiO_2 contents, in agreement with the calculated cavitation threshold. Their values are ~1/2 of the calculated cavitation threshold (700 to 800MPa) for these materials, also in accordance with the predictions made by Eq. (8). The total strain increases drastically at 20 wt% SiO2, which prevents further loading above 200 MPa.

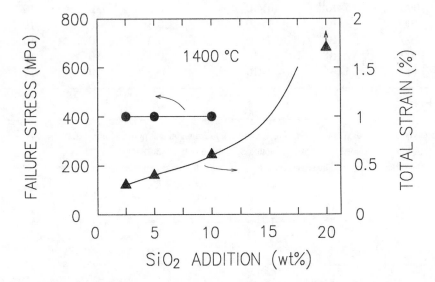

Figure 3 Delayed fracture strengths and total strains of Si_3N_4 with different amounts of SiO_2 in a stepwise loading test. The total strain is the accumulated strain when holding at 350 MPa was finished.

The microstructures of Si_3N_4 containing 10 and 20 wt% SiO_2 are shown in Fig. 4. With 10 wt% addition, the Si_3N_4 grains are continuously connected with only a thin intergranular layer of SiO_2 about 1 nm thick. By comparison, at 20 wt% SiO_2 addition, neighboring Si_3N_4 grains tend to be separated by a much thicker glassy phase. In other words, the formation of a continuous Si_3N_4–Si_3N_4 network is interrupted by the SiO_2 glass. These results show that the high fracture strengths of these ceramics, despite addition of up to 10 wt% SiO2, is probably due to the strong resistance of SiO_2 glass to cavitation. The considerable macroscopic creep ductility for ≥20 wt% SiO$_2$ addition can be attributed to the presence of thick glassy layers at the grain boundaries that allow macroscopic deformation by viscous flow of the glassy phase.

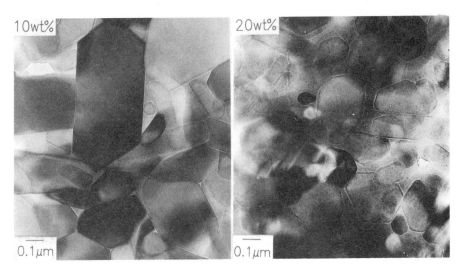

Figure 4 Transmission electron micrographs of Si$_3$N$_4$ with 10 and 20 wt% added SiO2.

Effect of Impurities

From Eq. (7) it is seen that the cavitation threshold depends on the distribution of impurities. The following exemplifies the case where only one kind of impurity is present. Here, defect i is defined as the defect–cluster composed of i single defects. To simplify the calculation, we suppose that the impurity is randomly distributed throughout the glassy phase. Its concentration (C_i) and local cavity nucleation energy decrease (ΔE_i) are related to those for a single defect by

$$C_i = C_1^{\ i} \tag{10}$$

and

$$\Delta E_i = i \Delta E_1 \tag{11}$$

where C_1 is the concentration and ΔE_1 is the local cavity nucleation energy decrease for a single defect.

The cavitation threshold calculated by Eq.(6) at 1400°C is given in Fig. 5 for different impurities (or ΔE_1) as a function of total impurity concentration. The resulting threshold stress is such that one tenth of glass pockets with a typical volume of (0.1 µm)^3 develop a cavity after undergoing a constant hydrostatic tension for 10^5 s. The other required physical parameters are given in Table 1. The condition of $\Delta G_{ic} > 0$ leads to a definite cavitation threshold for each value of impurity concentration. It can be seen from Figure 5, that the most important factor is ΔE_I, the decrease of local activation energy for cavity nucleation caused by a single defect. For $\Delta E_I = 500$ kJ/mol, even 1 ppm of such defects can decrease the intrinsic cavitation threshold of SiO_2 by ~ 35%. The total impurity concentration is also an important factor, because the higher this factor, the higher is the concentration of large defect-clusters.

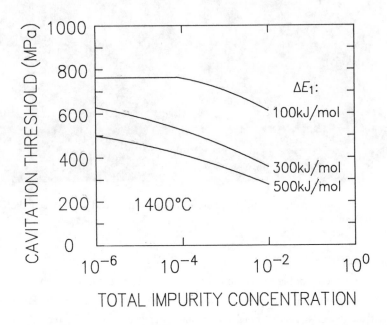

Figure 5 Cavitation threshold of SiO_2 at 1400°C for different ΔE_I and impurity concentrations.

The lifetimes of two dense silicon nitrides are shown in Fig. 6. The 10^5 s–failure stress of the high purity Si3N4 with 5 wt% SiO_2 is ~350 MPa at 1400°C. On the other hand, that of Si_3N_4 containing ~200 ppm fluorine, according to Hermansson et al. [21], is ~150 MPa even at 1100°C. The intergranular SiO_2 glass in this material is estimated to contain 1 % fluorine. The cavitation thresholds of the intergranular glassy phases in these two materials, calculated by using both the existing model and the present one, are shown in Fig. 7. The predictions given by the existing model are not in agreement with the measured results. However, the present model predicts a decrease proportional to the decrease in the delayed fracture strengths. The delayed fracture strengths of these ceramics are ~1/2 of the cavitation thresholds of their intergranular glassy phases, as predicted by Eq. (8). These results indicate that it is possible to predict the delayed fracture strength of Si_3N_4 from the physical properties of its intergranular glassy phases.

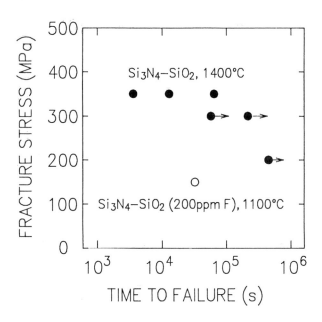

Figure 6 Lifetimes of a high–purity Si_3N_4 and a Si_3N_4 containing ~200 ppm fluorine.

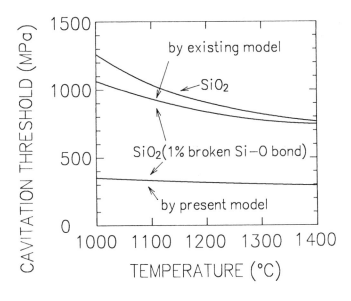

Figure 7 Temperature dependence of the cavitation thresholds of SiO_2 glass and of SiO_2 glass containing 1% fluorine impurity.

SUMMARY

This paper focuses on the relation between the delayed fracture strength of Si_3N_4 at high temperatures and the physical properties of the intergranular glass phase. The following conclusions can be drawn:

(1) Athermal defects are considered as cavitation sites associated with lower local activation energy for cavity nucleation. This model suggests a quantitative explanation of the detrimental effect of fluorine on the delayed fracture strength of Si_3N_4.

(2) The presence of ≤10 wt% SiO_2 is not detrimental to the high temperature strength of Si_3N_4. The high strength of Si_3N_4–SiO_2 ceramics can be explained by the high cavitation threshold of $SiO2$, which is independent of the volume of the glassy phase.

(3) The delayed fracture strength of Si_3N_4 was shown to be ~1/2 of the cavitation threshold of its intergranular glassy phase. The potential of predicting the lifetime of Si_3N_4 at high temperatures from the physical properties of the intergranular glass phase is demonstrated.

REFERENCES

1. Raj, R. and Ashby, M.F., Acta Metall., 1975, **23**, 653-66.
2. Evans, A.G., Rice, J.R., and Hirth, H.P., J. Am. Ceram. Soc., 1980, **63**, 368-75.
3. Tsai, R.L. and Raj, R., Acta Metall., 1982, **30**, 1043-58.
4. Marion, E., Evans, A.G., Drory, M.D., and Clarke, D.R., Acta Metall., 1984, **31**, 1445-57.
5. Thouless M.D. and Evans, A.G., J. Am. Ceram. Soc., 1984, **67**, 721-7.
6. Tanaka, I., Pezzotti, G., Matsushita, K., Miyamoto, Y., and Okamoto, T., J. Am. Ceram. Soc., 1991, **74**, 752-9.
7. Fisher, J.C., J. Appl. Phys., 1948, **19**, 1062-8.
8. Hirth, J.P. and Pound, G.M., Condensation and Evaporation-Nucleation and Growth Kinetics. Pergamon Press/MacMillan, New York, 1963.
9. Kingery, D., Bowen, H.K., and Uhlmann, D.R., Introduction to Ceramics, Second edition, John Wiley & Sons Inc., New York, 1960.
10. Nieh, T.G. and Nix, W.D., Scripta Metall., 1980, **14**, 365-8.
11. Shiozawa, K. and Weeertman, J.R., Acta Metall., 1983, **31**, 993-1004.
12. Chen, C.W. and Machlin, E.S., 1956, **4**, 655-6.
13. Watanabe, T. Metall. Trans., 1983, **14A**, 531-45.
14. Chan, K.S., Page, R.A., and Lankford, J., Acta Metall., 1986, **34**, 2361-70.
15. Lim, L. C., Acta Metall., 1987 **35** 1663-73.
16. Hsueh, C.H. and Evans, A.G., Acta Metall., 1981 **29** 1907-12.
17. Hetherington, G.K., Jack, H. and Kennedy, J.C., Phys. Chem. Glasses, 1964 **5** 130-6.
18. Kingery, W.D., J. Am. Ceram. Soc., 1959 **42** 6-10.
19. Zeng, J., Tanaka, I., Miyamoto, Y., Yamada, O., and Niihara, K., J. Am. Ceram. Soc., 1992 **75** 148-52.
20. Zeng, J., Tanaka, I., Miyamoto, Y., Yamada, O., and Niihara, K., J. Am. Ceram. Soc., 1992 **75** 195-200.
21. Hermansson, L., Burstrom, M., Johansson, T., and Hatcher, M., J. Am. Ceram. Soc., 1988 **71** C183-5.

FABRICATION, MICROSTRUCTURE, AND SOME PROPERTIES OF Si_3N_4-BASED NANOCOMPOSITES

TATSUJI MATSUI, AKIRA YAMAKAWA
Itami Research Laboratories,
Sumitomo Electric Industries,Ltd
1-1-1, Koya-kita, Itami, Hyogo, 664 Japan

KOICHI NIIHARA
The Institute for Scientific and Industrial Research,
Osaka University
Mihogaoka 8-1, Ibaraki, Osaka, 567 Japan

ABSTRACT

The fabrication process and microstructure of Si_3N_4-based nanocomposites were investigated for the improvement of fracture toughness. The fine Si_3N_4 powders containing metal nitride particles such as VN, NbN, TaN and TiN were synthesized by heating the mixtures of Si_3N_4 and corresponding oxide powders in a N_2 atmosphere. The obtained VN/Si_3N_4 and TiN/Si_3N_4 powders were confirmed to be very fine (20 to 30 nm). These composite powders were sintered with sintering aids. No improvement in fracture toughness was observed for the VN/Si_3N_4 composites. For the TiN/Si_3N_4 composites, in which nano-sized TiN grains were dispersed within the Si_3N_4 matrix grains, the fracture toughness was increased. This was probably due to crack propagation through intragrains of the Si_3N_4 matrix.

INTRODUCTION

Si_3N_4 ceramics have excellent properties such as good abrasion and corrosion resistance, heat-stability, and high fracture toughness and strength. Therefore, they are expected to be promising materials for structural parts of highly efficient gas turbine engines and other engineering machines. However, the fracture toughness of Si_3N_4 is not yet adequate for practical use in severe environments. To overcome this problem many studies on dispersing particles, whiskers, or fibers into Si_3N_4 ceramic materials have been carried out extensively to improve the toughness of Si_3N_4 ceramics [1-5]. Recently, Niihara et al. have demonstrated that some kinds of ceramics were highly toughened by the dispersion of nanometer-sized particles into the matrix grains [6-8]. Furthermore, it has been also shown that such nanocomposites have improved fracture strength and thermal stability as well as fracture toughness.

In this study, the fabrication process of Si_3N_4-based nanocomposites was investigated to obtain highly toughened Si_3N_4 ceramics. Metal nitrides, such as VN, NbN, TaN, and TiN were selected for the second phase dispersoids. The fine Si_3N_4 powder mixed with these metal nitrides was synthesized by heat-treating the corresponding metal oxide in a N_2 atmosphere and its sintering behavior was examined.

MATERIALS AND METHODS

Selection of dispersants
VN, NbN, TaN, and TiN were selected as the dispersants for the Si_3N_4 matrix. Because VN, NbN, and TiN have much larger coefficients of thermal expansion than Si_3N_4, the effects of residual stress caused by the difference of thermal expansion coefficients between the matrix and dispersant were expected to be significant. TaN was also studied for comparison, because of the similarity of its thermal expansion coefficient to Si_3N_4.

Synthesis of the composite powder
The composite powder of Si_3N_4 and metal nitride was synthesized via in-situ nitridation of the corresponding metal oxide, and not by the addition of metal nitride powder to Si_3N_4 powder. Each of the metal oxide powders, V_2O_5, Nb_2O_5, Ta_2O_5, and TiO_2 was separately mixed in ethanol with the Si_3N_4 powder. These mixtures were CIP'ed (cold isostatically pressed) under a pressure of 1.5 ton/cm^2, then heat-treated at 1500°C for 12 hours in a nitrogen atmosphere. The resulting powders were pulverized and studied by X-ray diffraction analysis (XRD) and transmission electron microscopy (TEM).

Consolidation of the synthesized powder
Y_2O_3 and Al_2O_3 (8 wt%) as sintering aids were added to the synthesized VN/Si_3N_4 and TiN/Si_3N_4 composite powders. After pressing and CIPing (1.5 ton/cm^2), the resulting mixtures were sintered at 1850°C for 1-6 hours in a nitrogen atmosphere. A commercial Si_3N_4 powder was also sintered under the same condition for comparison. Density, Vickers hardness, fracture toughness (K_{IC}, by indentation fracture method), and the crystal phases of the sintered bodies were estimated for all the sintered bodies. The microstructure was observed by TEM and scanning electron microscopy (SEM).

RESULTS AND DISCUSSION

Evaluation of the composite powder

TABLE 1 shows the crystal phases and average particle size of the composite powders obtained after heat-treatment. All the metal oxide starting materials, were converted to the corresponding metal nitride by heating at 1500°C for 12 hours. The VN/Si_3N_4 and TiN/Si_3N_4 mixtures were very fine powders with particle sizes of 20–30 nm, while NbN/Si_3N_4 and TaN/Si_3N_4 consisted of comparatively large particles of 100–500 nm, as shown in Figure 1.

TABLE 1
Crystal phases and average particle sizes of various
kinds of powder after heat-treatment at 1500°C for
12 hours in a nitrogen atmosphere

Starting mixture	Crystal phases	Average particle size (nm)
$Si_3N_4+V_2O_5$	Si_3N_4+VN	20–30
$Si_3N_4+Nb_2O_5$	Si_3N_4+NbN	500
$Si_3N_4+Ta_2O_5$	Si_3N_4+TaN	100–500
$Si_3N_4+TiO_2$	Si_3N_4+TiN	20–30

(a)VN/Si_3N_4 (b)NbN/Si_3N_4

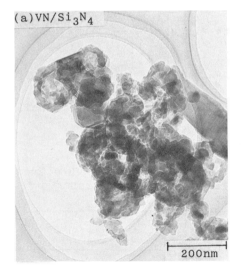
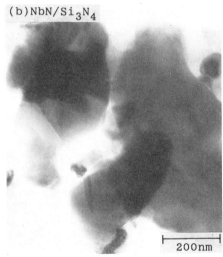

200nm   200nm

Figure 1. TEM photographs of the powders obtained by heat-treatment of the mixtures of (a)$V_2O_5+Si_3N_4$ and (b)$Nb_2O_5+Si_3N_4$ at 1500°C for 12 hours.

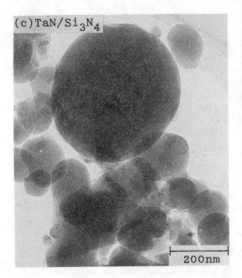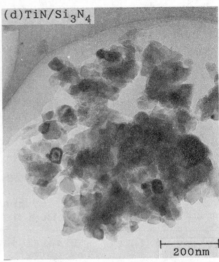

Figure 1. TEM photographs of the powders obtained by heat-treatment of the mixtures of (c)Ta_2O_5+Si_3N_4 and (d)TiO_2+Si_3N_4 at 1500°C for 12 hours.

Although the reason for these differences is not clear, it is suggested that the properties of the metal ions of the initial oxides affect the final morphology of the composite powder. The VN/Si_3N_4 and TiN/Si_3N_4 mixtures, which produced fine composite powders with nanosize particles, were used as the starting materials to fabricate the sintered bodies.

Evaluation of sintered bodies

VN/Si_3N_4 system: TABLE 2 shows the relative density, Vickers hardness, K_{IC}, and the VN average particle size of the sintered bodies made from the synthesized VN/Si_3N_4 composite powder and a commercial Si_3N_4 powder.

TABLE 2
Relative density(ρ), Vickers hardness(HV), K_{IC}, and VN average particle size(d) of VN(5vol%)/Si_3N_4 and Si_3N_4 sintered at 1850°C for 1 hour and 6 hours

Materials	Sintering time (hours)	ρ (%)	HV (kg/mm^2)	K_{IC} (MPa·m$^{1/2}$)	d (μm)
VN/Si_3N_4	1	95	1750	5.4	0.9
	6	96	1740	6.0	4.0
Si_3N_4	1	99	1890	5.2	——
	6	100	1850	6.1	——

All specimens were close to full density, but the composites, including 5vol%-VN, showed no significant increase of K_{IC}, and the hardness decreased compared with monolithic Si_3N_4 under similar conditions. In addition, the average particle size of the VN crystallites increased from 0.9 to 4.0 μm as the sintering time was increased from 1 to 6 hours. From these results, it is apparent that VN grains grow rapidly under the sintering conditions in this work and can not retain the nanosize of the starting powder after sintering. Thus, it seems to be very difficult to produce highly-toughened nanocomposites in the VN/Si_3N_4 system.

TiN/Si_3N_4 system: The sintered bodies of the TiN(5vol%)/Si_3N_4 composite powder consisted of elongated and equiaxed Si_3N_4 grains 0.3-3 μm (short axis) and 0.3-0.5 μm in size, respectively. Furthermore, some Si_3N_4 included the nanosized TiN dispersants within the grains, as shown in Figure 2. It has been thought that such microstructure was mainly caused by the use of a fine composite powder as a starting material, by the difficulty of TiN grain growth compared with Si_3N_4, and by crystallographic matching between TiN and Si_3N_4 grains. In the regions where some TiN dispersants were incorporated in the Si_3N_4 matrix grains, cracks generated by Vickers indentation on the surface of the specimen tended to propagate through the matrix grains transgranularly, and circumvent TiN grains dispersed within Si_3N_4 grains, as shown in Figure 3.

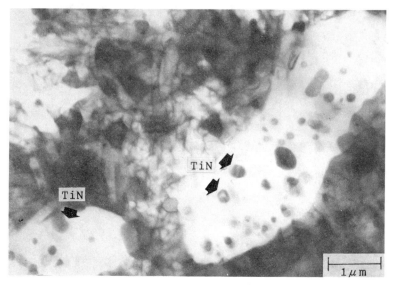

Figure 2. TEM photograph of microstructure of the TiN(5vol%)/Si_3N_4 composite obtained by sintering the synthesized powder at 1850°C for 2 hours.

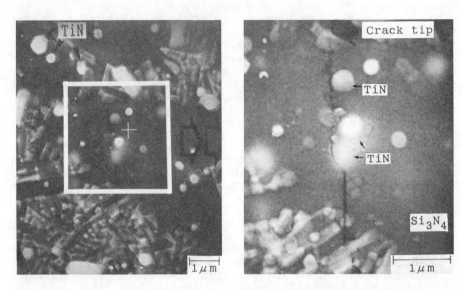

Figure 3. Crack propagation patterns in the region where some of the TiN dispersant was incorporated in the Si_3N_4 grains. (K_{IC} = 10.56 MPa·m$^{1/2}$)

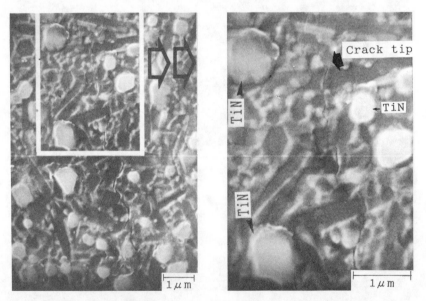

Figure 4. Crack propagation patterns in the region where most of the TiN particles were located outside the grains of Si_3N_4. (K_{IC} = 5.06 MPa·m$^{1/2}$)

Furthermore, the resistance against crack propagation was much higher in these regions (K_{IC} = 10.56 MPa·m$^{1/2}$, figure 3) than in the other regions where most of the TiN particles (about 1 μm in size) were located within the grain boundaries of the Si_3N_4 (K_{IC} = 5.06 MPa·m$^{1/2}$, figure 4).

These results suggest that both the position and the size of the dispersants in the matrix influence the patterns of crack propagation. Furthermore, the Si_3N_4 matrix can be toughened by small-sized dispersants with large coefficients of thermal expansion, such as TiN, incorporated in the matrix grains. Niihara has demonstrated that fracture mechanisms change from intergranular to transgranular fracture due to subgrain boundaries caused by the dispersion of nanosized SiC particles into an Al_2O_3 or MgO matrix, and that both toughness and strength have been much improved thereby [9,10]. With respect to the Si_3N_4 matrix, the effect of dispersants incorporated in the matrix grains on toughness is not yet adequately understood. But it seems clear that TiN particles located in the intragrains of Si_3N_4 interact with the crack tip more effectively than those located within the grain boundaries.

CONCLUSIONS

Composite powders of Si_3N_4 and various metal nitrides were prepared in order to fabricate Si_3N_4-based nanocomposites by a conventional powder metallurgy technique. For the VN/Si_3N_4 and TiN/Si_3N_4 composites, the powders obtained were found to be very useful for making Si_3N_4-based nanocomposites. In the VN/Si_3N_4 system, the VN grains showed rapid grain growth during the sintering process. No improvement in fracture toughness was observed. For the TiN/Si_3N_4 system, on the other hand, the nanosized TiN particles were mainly located within the matrix grains in most of the sintered body. Fracture toughness was strongly improved in those regions by the presence of the nanosized TiN dispersants, probably due to the propagation of the crack tip into the interior of the matrix grains. From these observations, it can be concluded that the TiN/Si_3N_4 nanocomposite system is very promising for developing highly tough ceramics.

REFERENCES

1. Niihara, K., Kogyo Zairyo, 1989, 37, 78-83.

2. Grekovich, C. and Palm, J.A., J. Amer. Ceram. Soc., 1980, 63, 597.

3. Buljan, S.T., Baldoni, J.G. and Huckabee, M.L., Ceram. Bull., 1987, 66, 347-352.

4. Izaki, K., Hakkei, K., Ando, K., Kawakami, T. and Niihara, K., In Ultrastructure Processing of Advanced Ceramics, ed. J.M. Mackenzie and D.R. Ulrich, John Wiley & Sons, Inc., New York, 1988, pp.891-900.

5. Matsui, T., Komura, O. and Miyake, M., Nippon Ceramics Kyokai Gakujutsu Ronbunshi, 1991, 99, 1103-1109.

6. Niihara, K. and Nakahira, A., Proc. of MRS Int. Meeting on Advanced Materials, 1989, 5, 129-134.

7. Niihara, K., Nakahira, A., Ueda, H. and Sasaki, H., Proc. of 1st Int. SAMPE Symp., Chiba, 1989, 1120-1125.

8. Niihara, K., Hirano, T., Nakahira, A., Ojima, K., Izaki, K. and Kawakami, T., Proc. of MRS Int. Meeting on Advanced Materials, Tokyo, 1988, pp.107-112.

9. Niihara, K., Nakahira, A. and Suganuma, K., New Ceramics, 1989, 2, 1-6.

10. Niihara, K., Ceramics, 1990, 25, 17-18.

APPENDIX

This work was mainly performed under the management of Engineering Research Association for High Performance Ceramics as a part of The Fine Ceramics R & D Project of Basic Technology for Future Industries supported by New Energy and Industrial Technology Development Organization.

Session V

Structural Ceramics III

GRAIN BOUNDARIES IN METALLIC MATERIALS

F. H. FROES, C. SURYANARAYANA and S. B. BHADURI
Institute for Materials and Advanced Processes (IMAP)
College of Mines and Earth Resources
University of Idaho
Moscow, Idaho 83843-4195, USA

ABSTRACT

The nature, characteristics, and properties of grain boundaries in different types of metallic materials are discussed, covering metals and intermetallics with normal grain sizes of $\geq 1 \mu m$ and also those with ultrafine grain sizes in the range of 5-20 nm. The effect that grain boundaries can have as sinks for vacancies is presented. Further, the importance of grain boundary solute segregation, its effect on mechanical behavior, and methods to minimize adverse effects are also discussed. Recent controversies regarding the differences in the nature of grain boundaries between nanocrystals and conventional materials are highlighted.

INTRODUCTION

Metals generally occur in the crystalline state with atoms in a periodically repeating geometric array in space [1]. In contrast this crystallinity is often not exhibited by many other materials such as wood, plastics, glass, and ceramics. The crystallinity of metals leads in turn to the major mechanism of plastic deformation in metals - the motion of line defects (dislocations). This behavior is a major reason for the mechanical characteristics exhibited by metals - moderate strength, high ductility, excellent fracture behavior, moderate elevated temperature performance, and moderate environmental resistance.

If all the atoms in a material continued with this periodic arrangement this would constitute a single crystal. In practice this situation rarely occurs - rather we have volumes of atoms (grains) with the periodicity mentioned above and other volumes of atoms with the same periodicity but differently oriented. The interfaces of these differently oriented regions are grain boundaries. Because many different orientations between adjacent grains are possible the nature of the grain boundaries can vary considerably [2]; for example in terms of the ease of dislocation passage from one grain to the next, and in the number of atoms in the grain boundary region which are randomly oriented rather than being in a periodic array. The overall behavior of metals is influenced both by the nature of the atoms themselves, their periodicity, the size of the grains, and the nature of the grain boundaries.

This paper will discuss grain boundaries in terminal alloy metals and intermetallics - with both conventional grain sizes ($\geq 1\mu$m) and those with ultrafine grain sizes in the range of 5-20 nm (1nm $= 10^{-9}$m).

GRAIN BOUNDARIES IN CONVENTIONAL METALS

Nature of Grain Boundaries: Generally grain boundaries have five degrees of freedom: three specifiying the orientation of one grain relative to the other and two specifying the orientation of the boundary relative to one of the grains [1]. The orientation mismatch can vary from quite small values (small-angle boundaries) to a large mismatch (high-angle boundaries). In the latter case a more open grain boundary region will exist with a larger number of grain boundary atoms not belonging to the crystal structure of either of the contacting grains. The grain boundary energy consists of two terms: one due to a strain effect (atom displacement to maintain bonds between atoms) and the other a chemical effect (due to the different bonding in the vicinity of the grain boundary compared to that occurring in the interior of the grains). Thus the grain boundary energy will vary quite considerably.

Recently the O-lattice description of grain boundaries looks for sites which can become the basis for matching regions [3,4]. This in turn leads to a determination of the periodicity of grain boundary regions and an indication of the characteristics of the boundary including porosity, solubility, diffusivity, energy and mobility.

Effect of Grain Boundaries on Dislocation Motion: The grain boundaries act to impede or arrest the motion of dislocations [5]. Thus a fine-grained material is stronger than a coarse-grained material, because of the increased grain boundary area. Specifically Hall and Petch [6,7] showed that $\sigma_y = \sigma_o + Kd^{-\frac{1}{2}}$ where σ_y is the yield stress, σ_o the lattice friction stress to move individual dislocations, K is a constant, and d the average grain size. This is basically an ambient temperature relationship in which glissile dislocation motion is the primary mechanism by which deformation occurs; at elevated temperatures other mechanisms such as dislocation climb and grain boundary sliding predominate [1, 2].

Grain Boundary Segregation: If a large solute atom is present within a matrix (e.g. Sn in Cu) there will be a driving force for this atom to migrate to an open grain boundary region to accommodate the lattice strain [1]. This can have a number of effects including (a) "pinning" of the grain boundary (see later) and (b) "hot shortness" in which a species such as S migrates to the grain boundary in steel producing a low melting point FeS phase which can cause problems during processing (a solution is to add Mn to form the higher melting point MnS phase).

Grain Shape: Two factors influence the shape of grains - the requirement to fill space, and to be in a state of (metastable) equilibrium particularly with regard to the total grain boundary surface area [1]. The net result is that three-dimensional grains are irregular polyhedrons with curved faces [8,9].

Migration of Grain Boundaries: There are two major reasons for the migration of grain boundaries, with the objective of reducing the overall free energy of the system: due to stored energy (deformation) and because of interface curvature [1]. The large number of dislocations present in deformed material increase the energy of the lattice, so that by recrystallizing (forming new grains which are virtually stored energy free) the overall free energy decreases [1]. The grain boundary velocity (v) in this case varies linerally with the amount of stored energy.

Surface stresses tend to hold an interface planar, thus when the interface is curved a pressure difference must be present. This is manifested as a chemical potential difference, with this potential being higher on the concave side of the interface. Thus atoms will tend to move to the convex side with the net result that the curved boundary moves in the direction of the concave region. Making various simplifying assumptions grain growth can be related to time (t) by $D = Kt^n$ where D is the grain size, K and n are constants, with n generally about ⅓ [10], except in high purity material at high temperatures where it is ½, as predicted from theory [11].

Four factors influence the migration rate of grain boundaries: impurity atoms, second phase particles or voids, temperature, and the orientation of adjacent grains [1,2]. As the temperature increases there is an exponential increase in diffusion rate and a corresponding large increase in mobility of the grain boundaries.

The large influence of impurity atoms and orientation of adjacent grains on mobility of grain boundaries was clearly demonstrated by Aust and Ritter [12] who showed that even very small amounts of Sn in Pb can greatly slow down boundary motion (Fig. 1). As little as 60 ppm Sn reduces the migration rate by a factor of 10^4. This effect is explained in terms of the dragging of the large impurity atoms (which prefer to be located at the less densely packed grain boundary regions rather than in the grain interior) by the grain boundary [13]. These workers also showed that particular boundaries in which there is good atomic fit, and therefore little solute segregation, are much more mobile than "random" grain boundaries.

Second phase particles or voids also restrain the motion of grain boundaries, with the pinning forces depending mainly on the size and number of particles [14-16].

Other effects: Three other effects of grain boundaries in metals will be considered: the role of grain boundaries in elevated temperature deformation, the amazing influence that extremely small amounts of trace elements can have on the ductility of normally brittle intermetallics, and the part that grain boundaries can play in controlling precipitation adjacent to grain boundaries.

Unlike the situation at ambient temperatures a major deformation mechanism at elevated temperatures is grain re-orientation [17]. Thus high strength/creep resistance is associated with a large grain size while high deformation levels, especially at low strain rates (i.e. superplastic behavior) occur at fine grain sizes.

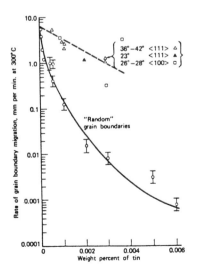

Figure 1. Influence of Sn impurity level on grain boundary migration rate in Pb[12].

The major effect that small amounts of trace elements can have on the mechanical behavior of intermetallics is most dramatically shown by the addition of B to Ni_3Al [18]. Single crystals of Ni_3Al are ductile at room temperature, but polycrystalline material fails by brittle grain boundary fracture with little plasticity. Small amounts of B (< 1000 ppm wt. %), added to Al-poor alloys, converted the polycrystalline Ni_3Al to a material with up to 50% elongation at room temperature, Fig. 2 [19]. It has been suggested that B enhances the ductility by either increasing the intrinsic strength or cohesion of the boundary, or it promotes generation of dislocations which aid in transmitting slip across the boundaries; however, the exact mechanism is not clear. This phenomenon is not unique to the Ni_3Al-B combination but does not appear to be wide-spread [19,20].

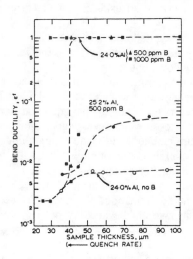

Figure 2. Influence of B in dramatically increasing the ductility of polycrystalline Al-poor polycrystalline Ni_3Al at ambient temperatures [19,20].

Grain boundaries in metals can also have a strong influence on precipitation of second phase particles in the vicinity of the grain boundary [21-23], Fig. 3 [22]. This occurs because of the ease of nucleation on the grain boundary itself resulting in grain boundary precipitates. It also takes place because grain boundaries act as a sink for vacancies during quenching from an elevated temperature solution annealing treatment. This latter effect gives rise to a precipitate free zone adjacent to the grain boundary because of reduced diffusion rates (a result of the lower vacancy content) during "pre-precipitate" aging and formation of nuclei below a critical size.

GRAIN BOUNDARIES IN NANOSTRUCTURED METALS

Nanostructured metals are single-phase or multi-phase polycrystals, the crystal size of which is of the order of a few (typically 1-10) nanometers in at least one dimension [24-28]. We will consider only nanostructure crystallites (three-dimensional nanostructures) in the present paper.

Nature of Grain Boundaries in Nanostructures: A schematic representation of a hard-sphere model of an equiaxed nanocrystalline metal is shown in Fig. 4. Two types of atoms are present: crystal atoms with nearest neighbor configurations corresponding to the

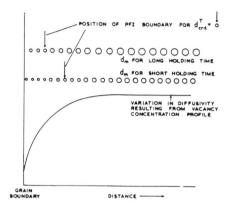

Figure 3.   Influence of grain boundary on the precipitation adjacent to this region. Vacancy profile and size of precipitate "nuclei" resulting from influence of vacancies on diffusion rate (left).   In this stainless steel both precipitation of carbides on the grain boundary itself and the reduced vacancy concentration close to the boundary have resulted in no fine carbides being present for a distance of about 1μm from the boundary and no stacking fault precipitation within about 5μm of the boundary (right) [22].

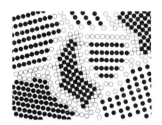

Figure 4.   Schematic representation of a hard-sphere model of an equiaxed nano-crystalline metal, ● atoms are located in a regular array within the grains, ○ atoms are not on a crystal lattice, and are loosely packed in the grain boundaries [24].

lattice and boundary atoms with a variety of interatomic spacings, differing from boundary to boundary.   A nanocrystalline metal contains typically a high number of interfaces ($\sim 6 \times 10^{19}$ cm^{-3}) with random orientation relationships and consequently a substantial fraction of atoms resides in the interfaces.   This can be as much as 50% for 5 nm grains and decrease to about 30% for 10 nm grains and 3% for 100 nm grains.

The grain boundaries in the nanocrystalline metals are relatively unrelaxed, a state somewhat similar to rapidly quenched metallic glasses [29] in that the system of boundary atoms has a local but not global energy minimum; thus, the material is in a metastable condition.

Since a large volume fraction of atoms resides in the grain boundaries in nanostructured materials, the structure of the interfaces can significantly affect their properties.   Employing various characterization techniques the grain boundary region has been interpreted in terms of a random structure [24].   This contrasts with the grain boundaries in normal metals which possess either short-range or long-range order.

High-resolution transmission electron microscopy investigations have been carried out recently on nanocrystalline palladium [30-32] with "wavy" grain boundaries exhibiting a qualitative difference in contrast observed in the vicinity of > 0.6 nm at the grain boundary. This was attributed to the high energy states of the grain boundaries in nanocrystalline materials, a situation not observed in conventional polycrystalline materials. In the nanometer-sized crystals ordering of the grain boundary structure to minimize the interfacial energy does not take place since the curvature of the grain boundary has a much bigger effect on the energy.

Other work [31,32] indicates that the grain boundaries in nanocrystalline materials are no different from those observed in conventional polycrystalline materials. By computer simulation methods it was shown that the nanocrystalline grain boundaries contain short-range ordered structural units representative of the bulk material suggesting that the atoms in nanocrystalline materials have sufficient mobility to rearrange themselves into low-energy configurations. Gleiter [24] points out that the very small thicknesses of high-resolution electron microscopy specimens and the high diffusivity in nanocrystalline materials may alter the atomic configuration at the grain boundaries, making it difficult to determine the structure of the boundaries accurately.

Results from field-ion microscopy and scanning tunneling microscopy are required to shed new light on the structure of the boundaries in the nanocrystalline materials.

Grain Growth: Grain growth studies in nanocrystalline materials are difficult since the grain size cannot be determined accurately [33-38]. In spite of the large interfacial volume (and increased grain boundary diffusivity) the nanometer-sized grains are surprisingly stable against growth, an effect which has been attributed to a narrow grain size distribution coupled with relatively flat grain boundary configurations; placing these nanophase structures in a local energy minimum, despite their large interfacial volume, and frustrating growth [39].

Lu [37] observed a $t^{1/2}$ time dependence for grain growth of Ni_3P nanocrystallities; however, in many instances the grain growth is rapid above a certain temperature and becomes negligible for longer annealing times. Since this transition coincides with the precipitation of a second phase the Zener drag mechanism is probably operative. However, in view of the extremely large driving forces available, other operative interfacial drag forces, e.g. triple junctions and pinning, may also occur [35]. Since the value of the exponent n during grain growth was found to be $< 1/2$, Höfler and Averback [33] concluded that pores present in the materials may also pin the grain boundaries and prevent grain growth. In a porosity-free material produced by sliding wear, the grain growth data could be fitted equally well to $n = 1/2$, $1/3$, or $1/4$ [34]. Thus, it was concluded that it is difficult to identify a grain growth mechanism on the basis of the exponent n alone, and that grain growth in nanocrystalline materials occurs in a manner similar to that in conventional polycrystalline materials.

Mechanical Properties: Nanocrystalline materials exhibit novel, and often, enhanced properties over those of conventional coarse-grained polycrystalline materials [24-28]. For example, a reduction in the grain size to the nanometer level often results in an increase in the strength and hardness along with an enhanced ductility in traditionally brittle ceramics and intermetallics. The increased strength has been interpreted in terms of the Hall-Petch relationship mentioned previously.

Hardness is plotted against $d^{-1/2}$ for nanocrystalline metals and alloys/compounds in Fig. 5 [40]. Although the hardness (or strength) is higher in comparison with coarse-grained polycrystalline materials, its variation with grain size in the nanometer regime does not always obey the Hall-Petch relationship, i.e., the slope K is sometimes positive, in other cases it is negative. This situation is exacerbated when it is realized that both positive and negative values of K have been reported by different investigators in the same grain size range, as demonstrated by the values for Pd. Further, TiO_2, Ni-P and Cu samples exhibit increasing hardness with decreasing grain size up to a critical value, below which the hardness decreases with decreasing grain size. The transition, however, occurs at 200 nm for TiO_2 (only a change in slope), 16 nm for Cu and at a much lower value (about 7 nm) for Ni-P alloys. One set of results on Ni-P shows a decreasing hardness with a decreasing grain size, while the other set shows a peak value at a grain size of 7 nm and a decreasing hardness on either side of this grain size. The absolute values also are quite different.

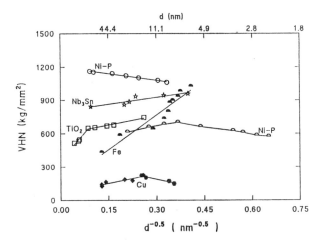

Figure 5.　　Hardness versus reciprocal square root of grain size for a number of nanocrystalline metals and alloys/compounds [40].

Possible reasons for the deviation from the Hall-Petch relationship in nanocrystalline materials can be proposed in terms of dislocation pile-ups. The Hall-Petch relationship was derived using the concept of dislocation pile-ups in individual grains [41], however, in very fine-grained nanostructured materials pile-ups cannot form when the grain size is less than a critical value ℓ_c (the dislocation spacing in the pile-up). Hence, Nieh and Wadsworth [42] suggested that when $d < \ell_c$, weakening mechanisms (e.g., viscous type flow) occur and lead to a decrease in hardness with decreasing grain size, i.e. a negative value for the slope K. The value of ℓ_c can be calculated by equating the repulsive force between the dislocations to the applied stress using the relation:

$$\ell_c = \frac{3Gb}{\pi(1-\nu)H}$$

where G is the shear modulus, b is the Burgers vector, ν is Poisson's ration, and H is the hardness.

This model has been only partially successful in explaining the results. A more convincing explanation for the negative value of K can be offered by taking into account the volume of triple junctions (i.e., intersection lines of three or more adjoining crystals) and the orientation of the grains [40]. From the available results, it appears that there is a critical grain size at which the nanocrystalline materials exhibit a peak in hardness and the hardness decreases at other grain sizes.

SUMMARY

The nature of grain boundaries in conventional and intermetallic metallic materials have been discussed. Grain boundaries can strongly influence the behavior of metals both directly and indirectly; the latter effect resulting from migration of solute or vacancies to these regions.

Nanostructured metals, with an extremely fine grain size (5-20 nm) can exhibit quite different behavior from conventional metals including high hardness, greatly enhanced ductility levels (especially under low strain rate conditions) and unexpectedly high levels of grain size stability.

ACKNOWLEDGEMENTS

The authors would like to express their appreciation to Mrs. Kandy Nelson for formatting and typing this paper.

REFERENCES

1.   Verhoeven, J. D., Fundamentals of Physical Metallurgy, John Wiley, New York, 1975.
2.   Porter, D. A. and Easterling, K. E., Phase Transformations in Metals and Alloys, Van Nostrand Reinhold (UK), Wokingham, Berkshire, 1983.
3.   Bollman, W., Crystal Defects and Crystalline Interfaces, Springer, Berlin, 1970.
4.   Ralph, B. and Randle, V., in Phase Transformations, ed. G. W. Lorimer, The Institute of Metals, London, UK, 1988, p. 618.
5.   Van Vlack, N., A Textbook of Materials Technology, Addison-Wesley, Reading, PA, 1973.
6.   Hall, E. O., Proc. Phys. Soc. London, 1951, **B64**, 747.
7.   Petch, N. J., JISI, 1953, **174**, 25.
8.   Smith, C. S., Metal Interfaces, ASM, New York, 1952, p. 65.
9.   Smith, C. S., Met. Reviews, 1954, **9**, No. 33, 1.
10.   Froes, F. H. et al., in Beta Titanium Alloys in the 1980's, eds. R. R. Boyer and H. W. Rosenberg, TMS-AIME, Warrendale, PA, 1984, p. 185.
11.   Hu, H. and Rath, B. B., Met. Trans., 1970, **1**, 3181.
12.   Aust, K. T. and Rutter, J. W., Trans. Met. Soc. AIME, 1959, **215**, 119.
13.   Hu, H., ed. The Nature and Behavior of Grain Boundaries, Plenum, NY, 1972.
14.   Zener, C. private communication to C. S. Smith, Trans. AIME, 1948, **175**, 17.
15.   Ashby, M. F., Harper, J. and Lewis, J., Trans. Met. Soc. AIME, 1969, **245**, 413.
16.   Froes, F. H., Cooke, C. M., Eylon, D. and Russell, K. C., in Sixth World Conference on Titanium, eds. P. Lacombe et al., Les Editions de Physique, Les Ulis Cedex, France, 1989, p. 1161.
17.   Coble, R. L., J. Appl. Phys., 1963, **34**, 1679.
18.   Aoki, K. and Izumi, O., J. Japan Inst. Met., 1979, **43**, 1190.

19. Koch, C. C., Horton, J. A., Liu, C. T., Cavin, O. B. and Scattergood, J. O., in Rapid Solidification Processing III, ed. R. Mehrabian, Nat. Bur. of Standards, Washington, D. C., 1983, p. 264.

20. Liu, C. T., Froes, F. H. and Stiegler, J. O., in vol. 2, 10th Edition of ASM International Metals Handbook, 1990, p. 913.

21. Kelly, A. and Nicholson, R. B., Prog. Met. Sci., 1963, 10, 149.

22. Froes, F. H., Ph.D. Thesis, "Precipitation of Carbides on Imperfections in Austenitic Steel", Sheffield University, Sheffield, England, 1967.

23. Froes, F. H., Wells, M. G. H. and Banerjee, B. R., Metal Sci. J., 1968, 2, 232.

24. Gleiter, H., Prog. Mater. Sci., 1989, 33, 223.

25. Suryanarayana, C. and Froes, F. H., in Physical Chemistry of Powder Metals Production and Processing, ed. W. Murray Small, TMS, Warrendale, PA, 1989, p. 279.

26. Suryanarayana, C. and Froes, F. H., Met. Trans. A, 1992, A23 (in press).

27. Siegel, R. W., Ann. Rev. Mater. Sci., 1991, 21, 559.

28. Siegel, R. W., in Processing of Metals and Alloys, vol. 15 of Materials Science & Technology - A Comprehensive Treatment, ed. R. W. Cahn, VCH Verlagsgesellschaft, Weinheim, Germany, 1991, p. 583.

29. Waseda, Y., The Structure of Non-Crystalline Materials, McGraw-Hill, New York, 1980.

30. Wunderlich, W., Ishida, Y. and Maurer, R., Scripta Met. Mater., 1990, 24, 403.

31. Thomas, G. J., Siegel, R. W. and Eastman, J. A., Mater. Res. Soc. Symp. Proc., 1989, 153, 13.

32. Thomas, G. J., Siegel, R. W. and Eastman, J. A., Scripta Met. Mater., 1990, 24, 201.

33. Höfler, H. J. and Averback, R. S., Scripta Met. Mater.,1990, 24, 2401.

34. Ganapathi, S. K., Owen, D. M. and Chokshi, A. H., Scripta Met. Mater., 1991, 25, 2699.

35. Boylan, K., Ostrander, D., Erb, U., Palumbo, G. and Aust, K. T., Scripta Met. Mater., 1991, 25, 2711.

36. Lu, K., Wei, W. D. and Wang, J. T., J. Appl. Phys., 1991, 69, 7345.

37. Lu, K., Scripta Met. Mater., 1991, 25, 2047.

38. Parker, J. C. and Siegel, R. W., J. Mater. Res., 1990, 5, 1246.

39. Siegel, R. W., in Superplasticity in Metals, Ceramics, and Intermetallics, eds. M. J. Mayo, J. Wadsworth, M. Kobayashi, and A. K. Mukherjee, Mater, Res. Soc., Pittsburgh, PA, 1990, vol. 196, p. 59.

40. Suryanarayana, C., Mukhopadhyay, D., Patankar, S. N. and Froes, F. H., J. Mater. Res. 1992, 7 (in press).

41. Armstrong, R. W., in Yield, Flow, and Fracture of Polycrystals, ed. T. N. Baker, Appl. Sci. Publ., London, England, 1983, p. 1.

42. Nieh, T. G. and Wadsworth, J., Scripta Met. Mater., 1991, 25, 955.

ENHANCEMENT OF TENSILE DUCTILITY IN NANOGRAIN SUPERPLASTIC CERAMICS THROUGH CONTROL OF INTERFACE CHEMISTRY

R. LAPPALAINEN, A. PANNIKKAT, AND R. RAJ
Department of Materials Science and Engineering
Cornell University
Ithaca, NY 14853-1501, USA

ABSTRACT

The strain rate and tensile ductility of superplastic nano-crystalline ceramics are controlled by the grain boundary cohesive strength and the rate of diffusion along interfaces. In this paper we propose that space charge layers at interfaces can enhance cohesion and ductility. We further propose that non-stoichiometric ceramics are more likely to exhibit such behavior. Data on magnesium aluminate spinels are presented to support this hypothesis. A model for superplastic flow in ceramics with charged layers at interfaces is presented. In addition we present a powder free route for the synthesis of nanocrystalline spinels, where transition metal dopants are employed to produce a uniform and ultrafine grain size.

INTRODUCTION

Superplastic deformation of ultrafine grain ceramics normally occurs by grain boundary sliding and diffusional accommodation of grain boundary sliding. For significant ductility to be achieved, it is preferable to have a ceramic microstructure composed of equiaxed grains with a grain boundary region resistant to decohesion and yet conductive for the diffusional transport of ions. In addition to the effect on the deformation mechanism, grain boundary properties have a significant effect on grain growth, which is an important issue because the effective viscosity of superplastic flow is very sensitive to the grain size; a twofold increase in grain size can produce an eightfold increase in the flow stress.

Dopants and impurities, that segregate to the grain boundaries, can influence superplastic behavior in different ways. The influence of very thin fluid layers has been extensively studied in the context of Si_3N_4/SiC composites [1], yttria-stabilized zirconia [2] and glass ceramics [3]. In this paper we consider atomistic mechanisms that can influence grain boundary properties in the absence of a fluid phase. Specifically, we consider situations where charged defects may lead to electrostatic effects that influence the superplastic mechanism where matter is transported from interfaces in compression to interfaces in tension by solid state diffusion. We propose that the electrostatic charge near the interface has two important consequences: (i) the space charge layer near the interface causes the stress/strain-rate response to become non-linear, and (ii) the electrostatic interaction across the interface can enhance the cohesive strength of the interface which can lead to greater tensile ductility in the ceramic. In this paper we present results on magnesium aluminate spinel that are consistent with these ideas. Being non-stoichiometric these spinels are more likely to contain charged defects at the interface.

It is noteworthy that very few finegrained ceramic materials, those that are free from an intergranular liquid phase, exhibit significant tensile ductility. In the literature, only yttria-stabilized zirconia, alloys of alumina and zirconia, hydroxyapatite and lead titanite [4-6] have been shown to be able to sustain a tensile strain greater than 10 % before failure. Interestingly, all these materials, like the spinel, are non-stoichiometric and therefore are likely to contain intrinsic and extrinsic charged defects. The segregation of these defects to the interfaces may have a bearing on the properties of the interfaces as discussed above.

In the experimental part of this work we present a new technique that has been developed to prepare and test free standing thin films of ultrafinegrained materials at high temperatures and variable strain rates. The films are prepared by (dual gun) electron beam evaporation which provides a powder free route for the preparation of specimens with homogeneous distribution of the constituents and gives great flexibility in the choice of the chemical composition. The as deposited structures are often amorphous, and the kinetics and the microstructure of crystallization can be controlled by the use of dopants that influence nucleation. In the present paper we show that the addition of platinum leads to a significant refinement of the grain size in the spinel.

EXPERIMENTAL

Tensile specimens were prepared using coevaporation with electron guns for deposition, and lithography for patterning as discussed in detail elsewhere [7,8]. High purity Al_2O_3, MgO, and magnesium aluminate spinel sintered pellets and platinum metal were used as source materials for evaporation. Two alumina rich spinel compositions $MgO:xAl_2O_3$, x=1.2 and 1.7 were evaporated. Another set of $MgO:1.25Al_2O_3$ samples was doped with 0.1-14 wt.% transition metal, Pt. During the deposition the vacuum was 0.1-1 mPa and the total deposition rate was 1 nm/s.

The deposition rates were measured with calibrated thickness monitors. The total length of the "dog-bone" shape specimens was 8 mm with a 4 mm long gage section and a 1x10, 50 microns cross section. In the same run, and under the same conditions, thin film samples were obtained for TEM and for crystallization studies.

The composition of the films was checked by Rutherford backscattering spectroscopy (RBS) and by the electron microprobe. Film thicknesses were determined with a profilometer, an ellipsometer, a scanning electron microscope and RBS. The thicknesses agreed within 5 % with the preset values during deposition. Microstructural features were studied using transmission electron microscopy and X-ray diffraction.

RESULTS

Tensile test results for spinels containing different platinum concentrations are reported. These samples consisted of spinel and 0, 0.1, 1, 8 and 14 wt.% Pt. The crystallinity and the grain size of the specimens was studied by TEM. The films were examined in the as deposited condition and after annealing and testing. The platinum content was found to have a significant effect on the crystallinity of the as deposited films (the deposition temperature of the films was expected to be less than 120°C); while the films made with 8 % and 14 % platinum were crystalline with a grain size of <10 nm, the films with 0, 0.1 % or 1 % platinum were amorphous. This suggests that the platinum has a role in crystal nucleation (this topic will be discussed in a separate paper [9]). The relative density of the as deposited films, as determined by RBS and ellipsometry, was about 75 % of the theoretical value.

Samples annealed between 1000-1300 °C, were all nanocrystalline. The 8 and 14 wt.% Pt samples consisted of Al_2MgO_4 spinel grains and precipitates, 10-50 nm in diameter. The composition of the precipitates was analyzed by STEM and determined to be primarily platinum. When annealed above 1000 °C the films rapidly sintered to a high density, presumably because of the nm scale grain size of the specimens. The final density of these specimens was measured to be greater than 94% of the theoretical value.

The grain size distributions of the 8 and 14 wt.% Pt/spinel samples were log-normal. In contrast, the samples with 0, 0.1 and 1 wt.% Pt had a bimodal distribution. The bimodal distribution consisted of nanocrystalline grains of spinel along with a few very large grains of α-alumina, ranging in size from 1.3 to 1.9 μm. The mean grain sizes of different fibers tested at 1200 °C and 1280 °C are given in Table 1. The grain size values given in Table 1 were obtained from TEM micrographs taken from the tested specimens and represent a mean value of the square root of area for about 200 grains.

Samples for mechanical testing were annealed in-situ before testing. The annealing procedure consisted of several stages. First, the samples were heated up to 800 °C with a

TABLE 1
The mean grain sizes for the fibers tested at 1200 and 1280 °C

material	d(nm) at 1200 °C	d(nm) at 1280 °C	abnormal grains at 1200 °C
spinel	99 *)	220 *)	1.3 μm ~ 30 vol.%
1 wt.% Pt/spinel	130 *)	160 *)	1.9 μm ~ 30 vol.%
8 wt.% Pt/spinel	67	88	-
14 wt.% Pt/spinel	91	210	-

*) the value doesn't include abnormal grains

heating rate of 15 °C/min and then with 2-10 °C/min up to 1100 °C. The specimens were kept without load at 1100 °C for 1 h and at the test temperature for a few hours before testing.

The stress-strain rate data for different compositions, all obtained at 1200°C, are shown in Fig. 1. These data were obtained by step-wise change of the strain rate in the range $5*10^{-7}$ to $5*10^{-4}$ s^{-1}. The effect of increasing platinum content is to increase the apparent stress exponent n from ~2 to ~4. The flow stress changes very little when the platinum content ranges from 0 - 8%, but decreases by a factor of more than twenty when the platinum content is increased to 14 %.

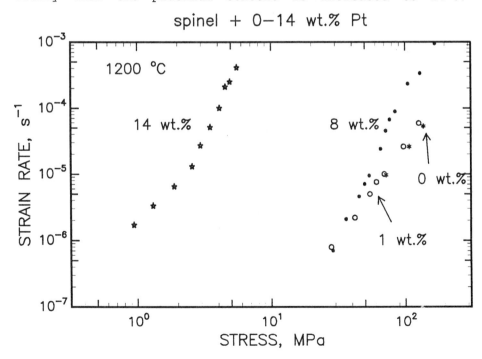

Figure 1. Stress-strain rate curves for 0, 1, 8 and 14 wt.% Pt/magnesium aluminate spinel samples at 1200 °C.

Since the grain size of the 8 % and the 14 % samples was about the same (see Table 1), this drop in the flow stress reflects a change in the grain boundary properties.

The 8 % and 14 % Pt specimens could be deformed up to about 35 % strain in tension before failure, while all the 0 - 1 % specimens failed before a tensile strain of ~10 %. This difference is attributed to the presence of the abnormally large alumina grains where cavities and cracks were observed to have initiated in the 0-1 % specimens. In the high platinum specimens the cavities were very small and were distributed throughout the specimen. Figure 2 shows the stress/strain curves for 1 and 8 wt.% Pt specimens which were deformed to fracture with a constant strain rate.

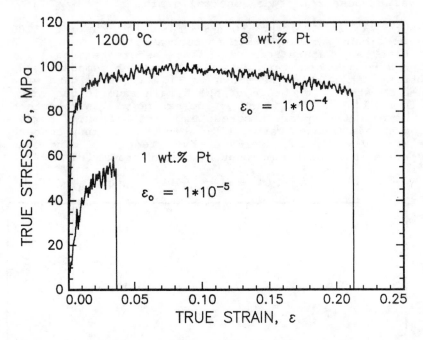

Figure 2. Stress-strain curves for 1 and 8 wt. % Pt specimens at 1200 °C. Notice a difference in strain rate.

The temperature dependence of the stress-strain curves for the 8 and 14 wt.% Pt specimens is shown in Fig. 3. The grain sizes of the samples tested at 1200 and 1280 °C are listed in Table 1. Even at the highest testing temperature (1280 °C) used in this study, only a few (< 1 vol.%) large abnormal grains existed in these specimens.

To get more insight into the deformation mechanism, the data were fitted to the following equation:

$$(\frac{\partial \varepsilon}{\partial t}) = \frac{A}{RT} \frac{\sigma^n}{d^p} \exp(-\frac{Q}{RT}) \tag{1}$$

Here n and p are stress and grain size exponents, Q is the activation energy and T is the absolute temperature. To

maintain a uniform microstructure in the films, they were
statically annealed at 1280 °C for 4 hours before testing at

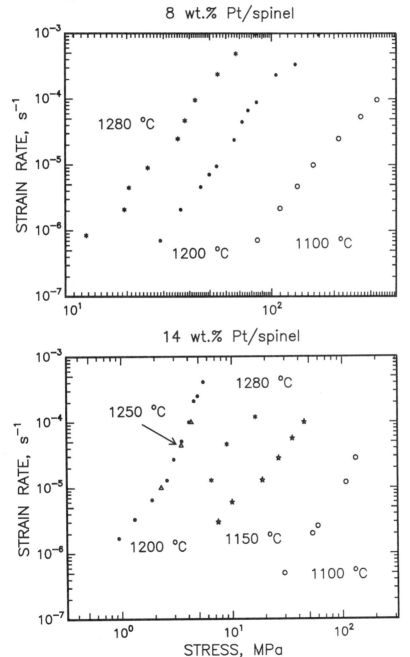

Figure 3. Stress-strain rate curves for 8 and 14 wt.%
Pt/spinel fibers at different temperatures.

temperatures in the range of 1100-1280 °C. The average activation energies obtained by fitting the equation (1) were 800 and 1000 kJ/mol for 8 and 14 wt.% Pt/spinel samples, respectively.

The grain size exponent p was measured by comparing the flow stress for samples of different grain sizes that were deformed at either 1100 °C or at 1200 °C. The 8 % Pt and the 14 % Pt specimens had a p value of 2.0 ± 0.3 when the strain rates were greater than 10^{-5} s^{-1}; at slower strain rates the value decreased to about 1. The values of the stress exponent n depended on the temperature and strain rate and were typically in the range 3-4.

DISCUSSION

The experimental results presented above can be summarized as follows: (i) platinum is an effective nucleating agent for grain refinement in magnesium aluminate spinels, (ii) the spinels exhibit a tensile ductility that is reminiscent of the deformation of zirconia alloys and hydroxyapatite, and (iii) the superplastic rheology cannot be adequately described by the classical equations for superplasticity that predict a linear dependence between the stress and the strain rate. Each of these points is explored in the following discussion.

The results have shown that the concentration of the platinum must be greater than 1 wt.% for it to serve as an effective nucleation agent in spinel. Such behavior is reminiscent of the processing of glass ceramics where a uniform density of nuclei, of the order of 10^{12} to $10^{15}/cm^3$, is obtained by the use of nucleating agents [10]. Nucleating agents such as platinum group metals are added to the glass ceramic batch during melting and nucleation is achieved with controlled heat treatment. Typical concentrations of these nucleating agents in glass ceramics are 4-12 wt.%. Thus, the work presented here raises the possibility of making fine grained ceramics by a powder free process, where the ceramic is first deposited in an amorphous form from the vapor phase (note that such materials may also be made by chemical vapor decomposition of metal-organic precursors), and then crystallized to obtain a dense and fine grained microstructure.

The platinum precipitates produced in the materials also serve a second purpose. They appear to restrict grain growth by the Zener pinning mechanism, which predicts the following limit for the grain size as a function of the size and the volume fraction of the pinning precipitates [11]:

$$d_l \approx d_i/f_i \tag{2}$$

Here d_l is the largest possible grain size, d_i is the size, and f_i the volume fraction of the precipitates that pin the grain boundaries. In our case, this simple estimate gives $d_l \approx 500$-1000 nm for 8 wt.% Pt/spinel which is significantly higher than the largest grains (< 150 nm) observed in tested fibers. It is possible that kinetic factors such as boundary mobility and diffusion at the platinum/spinel interface may have served to further restrain the rate of growth of spinel grains.

Next, we discuss the fact that non-stoichiometric oxides such as alloys of zirconia, and spinels, as demonstrated in the present work, exhibit unusual tensile ductility and greater resistance to intergranular fracture in superplastic deformation. We propose here that, as a result of asymmetrical segregation of charged species to the interfaces, the extrinsic and intrinsic charged defects in the non-stoichiometric oxides lead to space charge layers near the grain boundaries, and that the electrostatic potential created by this segregation influences diffusional flow processes required for superplastic flow.

As an example, we apply the above hypothesis to superplastic rheology in alloys of zirconia. The effects of the valence and the size of cation dopants have been studied in detail for tetragonal zirconia polycrystal (TZP) [12]. Only divalent and trivalent cations are enriched at the grain boundary and suppress grain growth, whereas higher valent cations are less effective. Since di and tri-valent impurity additions will create a greater concentration of point defects than tetravalent impurities, it would be expected that they will have a greater influence on diffusion and the energy of interfaces in zirconia. The effect of chemistry on the flow properties of yttria-stabilized zirconia has been studied, e.g. by Carry and Nauer [13]. The strain rate in Y-TZP depends strongly on the impurity content and is always higher for materials with higher impurity amounts. Furthermore, since the stress exponent has a tendency to decrease with increasing impurity content, superplasticity is enhanced by impurities. Even at low amounts, these impurities form amorphous phases at the grain boundaries (< 2 nm) of Y-TZP and affect the flow rate, especially at low stresses (< 10 Mpa).

The size of Pt^{2+} and Pt^{4+} cations (0.52 and 0.55 Å) is about the same as Al^{3+} (0.57 Å) and smaller than the size of Mg^{2+} cations (0.78 Å). Although the open crystal structure of spinel allows the high solubility for even large cations in contrast to e.g. alumina, platinum should not be very effective in suppressing grain growth based on the size and valence considerations only.

In the case of alumina rich magnesium aluminate spinel studied here, an extra amount of Al_2O_3 is expected to be in solid solution with $MgAl_2O_4$ spinel phase based on phase diagrams. The deformation rate and the concentration of defects should depend somewhat on the stoichiometry of the spinel. However, in this study, with two spinel compositions, x=1.2 and 1.7, the flow stress was found to increase with increasing alumina content which was mainly due to the difference in grain size distributions [9]. In the case of platinum doped spinel samples, the stoichiometry had only a minor effect on the flow behavior [9].

In the case of Y-TZP, the highest values of the stress exponent (n~3) were observed for fine grain and pure materials in the intermediate stress range [13]. Generally, n-values decreased (from 2.3 to 1.4) and p-values increased (from 1 to 3) with increasing strain rate. This tendency agrees well with the similar results obtained in this study. We assume that this behavior can be described by the following model.

We now present a phenomenological model for describing the effect of space charge on the diffusional transport of atomic species from one interface to another. The description is analogous to that of an electrochemical cell, except in the present case the driving force for the transport of ions is provided by the difference in the normal traction at interfaces. Thus diffusional transport required for superplastic deformation can be visualized as an electrochemical cell, the size of grain where ions are transported across the grain under a potential difference that is defined by the magnitudes of the compressive and tensile tractions on the opposite sides of the grain, and where space charges exist and form double layers at these grain interfaces due to the segregation of charged defects, as discussed earlier.

In the model we assume that the space charge layer creates a barrier for the exchange of ions with the interface and that the potential of this space charge layer is V_o. This potential opposes the chemical potential gradient that is provided by the applied stress and is equal to $\sigma\Omega/N_A$ where σ is the applied stress, Ω is the atomic volume and N_A is the Avogadro constant. Then the concentration of vacancies near to the surface where the compressive stress is acting is given by:

$$C_v = \exp\left(-\frac{E_f + (\sigma\Omega/N_A - qV_o)}{kT}\right) = C_{v,o}\exp\left(-\frac{\sigma\Omega/N_A - qV_o}{kT}\right) \qquad (3)$$

Here E_f is the formation energy and q is the charge of a vacancy, $C_{v,o}$ is the vacancy concentration under thermal equilibrium, k is the Boltzman constant and T is the absolute temperature.

As a first approximation, we can assume that the exponent in Eq. (3) is very much smaller than unity in which case that equation will lead to a simple modification of the Herring/Coble equation for diffusion flow, as follows:

$$\left(\frac{\partial\varepsilon}{\partial t}\right) = A\frac{(\sigma\Omega/N_A - qV_o)}{kT}\left[\frac{D_1}{d^2} + \frac{D_i\delta}{d^3}\right] \qquad (4)$$

where D_1 is the lattice diffusion and D_i is the interface diffusion and δ is the boundary width. The first term on the right is the contribution made by lattice diffusion and the second arises from boundary or interface diffusion; A is a dimensionless constant. The approximation that we have made here predicts a sharp threshold stress for deformation. In practice the transition across the threshold may be gradual, and this will be dealt with in more detail in a follow-up paper.

ACKNOWLEDGEMENTS

This research was supported by the Army Research Office under contract No: DAAL03-89-K-0132. Support was also received from the National Science Foundation through the use of the facilities of the Materials Science Center and the use of the processing equipment at the National Nanofabrication Facility, under grant No. ECS-8619049 at Cornell University.

REFERENCES

1. Wakai, F., Kodama, Y., Sakaguchi, S., Murayama, N., Izaki, K. and Niihara, K., _Nature_, 1990, **344**(6265), 421-423.

2. Nieh, T.G. and Wadsworth, _J., Acta Met. Mater._, 1990, **38**, 1121-1133.

3. Wang, J.-G. and Raj, R., _J. Amer. Ceram. Soc._, 1984, **67**(6), 399-409.

4. Wakai, F., _Br. Ceram. Trans. J._, 1989, **88**(6), 205-208.

5. Wakai, F. and Kato, H., _Adv. Ceram. Mater._, 1988, **3**(1), 71-76.

6. Wakai, F., Kodama, Y., Murayama, N., Sakaguchi, S., Rouxel, T., Sato, S. and Nonami, T., in _Superplasticity in Advanced Materials_, eds. S. Hori, M. Tokizane and N. Furushiro, The Japan Soc. for Research on Superplasticity, Osaka, 1991, p. 205-214.

7. Lappalainen, R. and Raj, R., _Acta Met. Mater._, 1991, _39_, 3125-3132.

8. Lappalainen R. and Raj, R., in _Microcomposites and Nanophase Materials_, ed. D. C. Van Aken, G. S. Was and A. K. Ghosh, TMS, Warrendale, PA, USA, 1991, pp. 41-51.

9. Lappalainen, R., Pannikkat, A. and Raj, R., to be submitted to Acta Metall.

10. McMillan, P. W., _Glass-ceramics_, Academic Press, New York, 1979, pp. 61-96.

11. Zener, C., see Smith, C. S., _Trans. AIME_, 1948, **175**, 15-51.

12. Chen, I.-W. and Liang, A. X, _J. Am. Ceam. Soc._, 1990, **73**(9), 2585-2609.

13. Carry, C., _Mat. Res. Soc. Symp. Proc._, 1990, **196**, 313-323.

DEVELOPMENT OF SUPERPLASTIC FUNCTIONAL CERAMICS THROUGH GRAIN REFINEMENT

F. WAKAI, Y. KODAMA, S. SAKAGUCHI, N. MURAYAMA, T. ROUXEL, N. SATO*
AND T. NONAMI*
Ceramic Science Department, Government Industrial Research Institute, Nagoya
1-1, Hirate, Kita, Nagoya 462 Japan
*Basic Materials Laboratory, TDK. Co. Ltd.
570-2, Matsugashita, Minamihatori, Narita, 286 Japan

ABSTRACT

Functional ceramics include those ceramics that have characteristic properties such as electronic, magnetic, optical, chemical, or biological properties. Superplasticity of the bioceramic, hydroxyapatite, may be the first example of superplastic elongation in a functional ceramic. Superplasticity in ZrO_2 can also be utilized as a unique forming method for the solid electrolyte for fuel cells. In this paper, the superplasticity of piezoelectric $PbTiO_3$ (lead titanate) is studied.

INTRODUCTION

Recent advances in superplastic ceramics have clearly demonstrated that superplasticity is one mode of deformation for fine-grained polycrystalline solids at elevated temperatures including metals [1], ionic polycrystals [2], and covalent polycrystals [3]. The phenomenon of superplasticity is characterized by exceptionally large elongation in tensile deformation. It is now the established concept that many ultrafine-grained ceramics can be superplastic when stretched slowly just below the sintering temperature.[4-8]

Superplasticity provides promising opportunities for the technology of ceramics that have useful properties, for example, electronic properties, magnetic properties, optical properties, biological properties and mechanical properties. Table I summarizes the key functions of ceramics and potentially superplastic materials. The superplasticity of structural ceramics such as Y_2O_3-stabilized tetragonal ZrO_2 (Y-TZP), Al_2O_3, mullite, Si_3N_4, SiC and their composites have already attracted much interest because of the potential for forming near-net shape components for mechanical applications. The superplasticity of fine-grained ceramics with properties other than mechanical properties also has been observed recently

It should be noted that a large number of electronic ceramics belong to the perovskite family. For example, we have demonstrated the superplastic elongation of $PbTiO_3$, which has a perovskite structure [9]. Furthermore superplastic ZrO_2 is also known to be a good oxygen ion conductor, and is used as a solid electrolyte. Hydroxyapatite, which is an example of so-called bioceramics, is also superplastic [10]. In this paper, the superplastic properties of

TABLE 1
Applications of ceramics

Field	Function	Materials	Device
Mechanical properties	Strength	Si_3N_4, SiC ZrO_2, Mullite, Al_2O_3	Components for engine
Electronic properties	Dielectric Piezoelectric Pyroelectric Insulation Semiconductive Superconductive Ionic conductive	$BaTiO_3$ PZT, $PbTiO_3$ PZT Al_2O_3 $BaTiO_3$ $YBa_2Cu_3O_{7-x}$, $Bi_2Sr_2Ca_2Cu_3O_{10}$ ZrO_2	Capacitor Transducer Infrared ray sensor IC substrate Thermistor Electric cable Fuel cell, oxygen sensor
Magnetic properties	Ferrimagnetic (Soft) (Hard)	$Zn_{1-x}Mn_xFe_2O_4$ $SrO·6Fe_2O_3$	Magnetic head Magnet
Optical properties	Transparent Electro-optic	ZnS PLZT	Laser window Optical switch
Biological properties	Bioactive Bioinert	Hydroxyapatite Al_2O_3, ZrO_2	Artificial bone

$PbTiO_3$ are studied as an example of superplasticity in functional ceramics. The materials underlined in Table 1 have been shown to be superplastic or highly ductile.

PIEZOELECTRIC MATERIAL - $PbTiO_3$

Ferroelectric materials such as PZT (55% $PbZrO_3$, 45% $PbTiO_3$) and $PbTiO_3$ are piezoelectrically active. They can be used for electromechanical transducers (e.g., ultrasonic transducers, vibration sensors, surface acoustic wave devices). Piezoelectric ceramics are formed into various shapes for device applications. In this paper, the deformation of lead titanate ($PbTiO_3$) with a perovskite structure is studied to ascertain its superplastic elongation.

Experimental

The material was sintered at 1240 °C for 5 hours in an O_2 atmosphere (1 atm). The density (7.15 g/cm^3) was almost equal to the theoretical density. The grains were equiaxed and 2.8 µm in diameter. Tension specimens were cut from sintered disks (diameter 78 mm, thickness 15 mm). The specimen had a gauge length of 11 mm with a circular cross section about 3 mm in diameter. The tension test was conducted in air. The cross-head speed of the universal test machine was controlled by a computer so that the tension test was conducted at a constant strain rate. A jump-strain rate test was also performed by changing the strain rate at each true strain of 0.1.

Results

The experimental results are summarized in Table 2.

TABLE 2
Summary of tension test data for PbTiO₃

Material	Temperature (°C)	Strain rate (s⁻¹)	Elongation (%)	Yield stress (MPa)
PT1	1050	Jump 1×10^{-4}, 5×10^{-5}	100	4.0
PT2	1100	Jump 1×10^{-4}, 5×10^{-5}	153	2.1
PT3	1150	Jump 1×10^{-4}, 5×10^{-5}	170	1.3
PT4	1150	Jump 1×10^{-3}, 5×10^{-4}	16	2.9
PT5	950	1×10^{-4}	5	48

The true stress-true strain curves in the jump-strain rate test are shown in Fig. 1. A superplastic elongation larger than 100 % could be observed at temperatures from 1050 °C to 1150 °C. The periodical drop of flow stress in the true stress-true strain curves in Fig. 1 corresponds to the change of true strain rate from 1×10^{-4} s⁻¹ to 5×10^{-5} s⁻¹. Generally the flow stress decreased with increasing temperature. The flow stress (below 2 MPa) of PbTiO₃ at 1150 °C was considerably lower than the flow stress of superplastic Y-TZP at the same test condition. An upper yield point at a strain of 5 % and subsequent strain hardening was observed at 1050 °C. On the other hand, the amount of strain hardening after the peak stress at 5 % strain was small at 1100 °C. An upper yield point was not observed at 1150 °C, and the stress increased gradually until the true strain of 0.7, then decreased slightly. The specimen tested at 950 °C fractured at 48 MPa and an elongation of only 5 %.

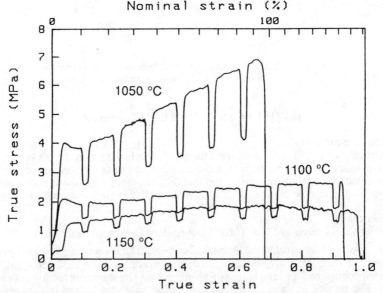

Figure 1. True stress-true strain curves of PbTiO₃ in a jump-strain rate test (1×10^{-4} s⁻¹, 5×10^{-5} s⁻¹):

When we assume the following relation between the flow stress and strain rate,

$$\dot{\varepsilon} = A \, \sigma^n \exp(-Q/RT) \tag{1}$$

The stress exponent (n) is calculated from the flow stresses (σ_1 and σ_2), which correspond to strain rate ($\dot{\varepsilon}_1$ and $\dot{\varepsilon}_2$),

$$n = \log(\dot{\varepsilon}_1/\dot{\varepsilon}_2)/\log(\sigma_1/\sigma_2) \tag{2}$$

The flow stresses at two strain rates were read from the curves in Fig. 1, and the apparent stress exponents derived from Eq. (2) were plotted in Fig. 2 as a function of strain. The initial stress exponents at 1050 °C and 1100 °C were 1.7 and 2.2 respectively, and they decreased gradually with increasing strain. On the other hand, the initial stress exponent of 2.2 at 1150 °C increased up to about 2.7 with increasing strain. The stress exponent calculated from the yield stresses of two specimens tested at 1150 °C (ZP2 and ZP5) was 2.7, and was identical with the results from the jump-strain rate test.

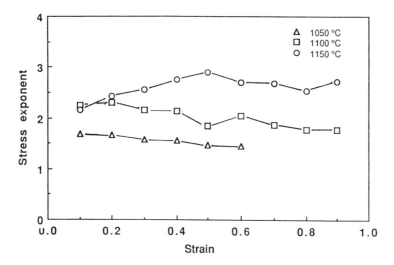

Fig. 2 Stress exponent of PbTiO$_3$ as a function of strain.

The flow stress at 1×10^{-4} s^{-1} is plotted as a function of temperature in Fig. 3. The calculated activation energy (-Q/n) was -182 kJ/mol in the temperature region higher than 1050 °C, and -333 kJ/mol in the temperature region from 950 to 1050 °C.

The ratio of the cross section to the initial cross section is plotted as a function of gauge position in Fig. 4. The tension specimen elongated at 1050 °C deformed uniformly just like other superplastic ceramics such as Y-TZP. But diffuse necking was observed for the specimens that were deformed at 1100 °C and 1150 °C.

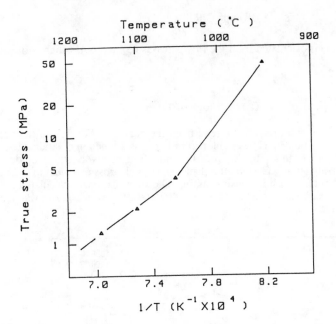

Fig. 3 Yield stress of PbTiO$_3$ as a function of temperature at 1 X 10^{-4} s^{-1}.

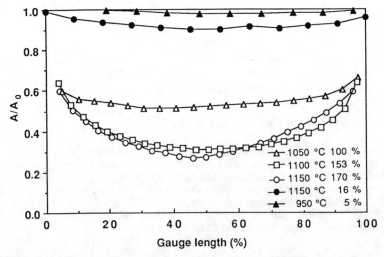

Fig. 4 Ratio of the cross section to the initial cross section of PbTiO$_3$ specimen as a function of gauge position.

The general condition for plastic stability in tension is;

$$m + \gamma \geq 1 \qquad (3)$$

where m is the strain rate sensitivity index and γ is an exponent for strain hardening. The m-value decreased with increasing temperature because the stress exponent increased up to 2.7 at 1150 °C. The smaller value of m could explain the diffuse necking at elevated temperatures. The strain hardening at 1050 °C could be beneficial for uniform deformation.

It is clear that several deformation mechanisms are operating during the superplastic flow of $PbTiO_3$ because the stress exponents and the activation energies are dependent on temperature. Further study to clarify the mechanisms is now in progress.

CONCLUDING REMARK

Recent research is rapidly uncovering superplasticity in a wide variety of fine-grained ceramics (e.g., structural ceramics, composites, electronic ceramics, bioceramics) If the properties of ceramics can be improved substantially by reducing the grain size to the nanometer level, the deformation processing of such fine-grained ceramics by utilizing superplasticity should find niche applications in the future in the ceramic industry.

ACKNOWLEDGMENTS

Part of this work was presented at the International Conference on Superplasticity in Advanced Materials (ICSAM-91) which was held on June 3-6, 1991 in Osaka, Japan.

REFERENCES

1. Pearson, C. E. , J. Inst. Metals, 1934, **54** , 111.

2. Wakai, F., Sakaguchi, S. and Matsuno, Y., Superplasticity of Yttria-Stabilized Tetragonal ZrO_2 Polycrystals, Advanced Ceramic Materials, 1986, **1** [3] 259-63.

3. Wakai, F., Kodama, Y., Sakaguchi, S., Murayama, N., Izaki, K. and Niihara, K., A Superplastic Covalent Crystal Composite, Nature, 1990, **344** [6265] 421-423.

4. Wakai, F., A Review of Superplasticity in ZrO_2-Toughened Ceramics, British Ceramic Transactions and Journal, 1989, **88** [6] 205-208.

5. Maehara, Y. and Langdon, T.G., J. Mater. Sci., 1990, **25**, 2275.

6. Chen, I.W. and Xue, L.A., J. Am. Ceram. Soc., 1990, **73**, 2585.

7. Wakai, F., Superplasticity of Ceramics, Ceramics International, 1991, **17** [3] 153-163 .

8. Nieh, T.G., Wadsworth, J. and Wakai, F., Recent Advances in Superplastic Ceramics and Ceramic Composites, International Materials Review, 1991, **36** [4] 146-161.

9. Wakai, F., Kodama, Y., Murayama, N., Sakaguchi, S., Rouxel, T., Sato, N., and Nonami, T., Superplasticity of Functional Ceramics, in Superplasticity in Advanced Materials, ed. S. Hori, M. Tokizane and N. Furushiro, The Japan Society for Research on Superplasticity, 1991, pp. 205-214.

10. Wakai, F., Kodama, Y., Sakaguchi, S. and Nonami, T., Superplasticity of Hot-Isostatically Pressed Hydroxyapatite, J. Am. Ceram. Soc., 1990, **73** [2] 457-60.

SOLUTE DRAG ON GRAIN BOUNDARY IN IONIC SOLIDS—THE SPACE CHARGE EFFECT†

I-WEI CHEN
Department of Materials Science and Engineering
University of Michigan
Ann Arbor, Michigan 48109-2136

ABSTRACT

Solute drag involving space charge interactions between aliovalent dopants and counterions near a grain boundary is evaluated. We use a lattice model to describe site competition between solutes and host ions, which naturally leads to saturated solute segregation at the grain boundary at higher concentrations. In such a limit, the screening distance is nearly concentration independent, and the grain boundary mobility goes through a minimum at an intermediate concentration. In the dilute limit, the mobility decreases monotonically with solute concentration, and a slower diffusing cation dopant of a larger effective charge provides a larger drag. Recent grain growth data of ZrO_2 and CeO_2 are in support of the above model.

INTRODUCTION

The interactions of diffusing impurities with grain boundaries have received considerable attention in ceramics and metallurgy.[1-4] They cause solutes to segregate or desegregate to a stationary grain boundary. Subsequent grain boundary movement will necessitate dragging the solute cloud along with it. The resultant "frictional" force on the boundary is dependent on the extent of segregation, the diffusivity of solutes, and the velocity of the grain boundary. This problem has been analyzed first by Cahn,[3] and more recently and rigorously by Srolovitz et al.[5] In particular, elastic interactions due to a size misfit or a modulus misfit have been considered. These classical potentials, which can be expressed as unique functions of the distance from the grain boundary, are independent of the velocity of the grain boundary or the distribution of other solutes. The principal application of this

---

† Supported by U.S. Department of Energy (BES) under Grant No. DE-FG02-87ER45302.

model is on grain size control, which can be achieved by lowering grain boundary mobility.

Although the above concept has been widely used in ceramic literature, the major interactions here involve not only elastic interactions but also space charge interaction. Unlike elastic interactions, however, the space charge interaction is dependent on the distribution of other solutes. The electrostatic potential is thus concentration dependent. In particular, the range of the interaction, dictated by the Debye-Hückel screening distance, decreases with solute concentration. The interaction is also velocity dependent in that the redistribution of the charged solute, in response to the movement of the grain boundary, modifies the interaction at the same time. Therefore, the analogy with the classical solute drag problem, as was assumed by previous investigations in the field,[1] is not valid. In this paper, we re-analyze this problem using fluorite-structured oxide as a reference model.

In Section II, a simple thermodynamic model of an electrolytic solid solution is presented. The solution of the solute concentration profile near a stationary grain boundary is obtained. In Section III, we formulate the diffusion equation for solutes, then consider the frictional forces on them by developing a force equation and a dissipation theorem. In Section IV, the mobility of a moving grain boundary is evaluated approximately by invoking the dissipation theorem and using the static concentration solution. Limitations and justifications of our model are further discussed in Section V. The major new predictions on grain boundary mobility, some of them quite unexpected at first sight, are: a much weaker concentration dependence in the dilute concentration range, a minimum in the intermediate concentration range, and a reversal of the dependence on the effective charge of the solute from the dilute to the concentrated range. Experimental verification of these predictions are mentioned briefly in Section V but its details are reported elsewhere.

THERMODYNAMICS OF AN ELECTROLYTIC SOLID SOLUTION WITH A STATIONARY GRAIN BOUNDARY

Solute and Defect Equilibrium

Consider a substitutional solid solution containing a host cation A, an anion B and an accepter cation solute X. Specifically, we have in mind a fluorite-structured oxide such as ZrO_2 or CeO_2 with divalent (e.g., Mg^{2+}, Ca^{2+}) and trivalent (e.g., Sc^{3+}, Y^{3+}, Gd^{3+}) cation dopants. (We shall refer to the above dopants as acceptor dopants.) The host compound is written as $A_\alpha B_\beta$. Using Vink-Kroger notation to denote defects, we let the dominant defect associated with the dopant be X_A with an effective charge $-z$. We also let the dominant lattice defect be V_B with an effective charge z'. We assume that if the molar fraction of cation substitution, c_0, in the bulk is much higher than the fraction of thermally produced charged defects in an undoped (pure) ionic solid, i.e., it is in the extrinsic regime, then the molar fraction of vacant anion sites, c_0', in the bulk must obey

$$\alpha z c_0 = \beta z' c_0' \tag{1}$$

to maintain charge compensation.

Now consider an interaction which causes an increase in the energy of cation substitution by U. Then, with a lattice model for site occupancy by solute ions in mind, we can describe the local fraction c of the dopant by the following "Law of Mass Action"[6]

$$c / (1-c) = (c_0 / (1-c_0)) \exp(-U/kT) \tag{2}$$

Rearranging the above, we obtain

$$c = 1 / [1 + \exp(U/kT) (1-c_0)/c_0] \tag{3}$$

The above expression for solute fraction reduces to the Boltzmann distribution, $c = c_0 \exp(-U/kT)$, in the limit of dilute concentration. At higher concentrations, competition of lattice sites dictates that saturation occurs with c approaching unity.[6] This form for solute (defect) distribution is identical to the Fermi-Dirac distribution in statistical physics for fermions. A similar set of relations holds for anion vacancies if we replace c_0 by c_0', c by c' and U by U', with c' and U' being the molar fraction and potential appropriate for anion vacancies.

Space Charge Problem

Near a geometric discontinuity, such as a grain boundary, an electrostatic potential exists in an ionic solid, as first pointed out by Frenkel[7] and later analyzed by Eshelby, Koehler and their coworkers.[8-9] The electrostatic potential is determined by solving Poisson's equation

$$d^2\phi/ds^2 = ne (\alpha z c - \beta z' c') / \varepsilon \tag{4}$$

where ϕ is the electrostatic potential, s is the distance from a plannar grain boundary, n is the concentration of the $A_\alpha B_\beta$ formula units, and ε is the dielectric constant. This potential is responsible for the interaction which the charged solute and defect species experience in the solid. For cation solutes, $U = -ze\phi$. For anion vacancies, $U' = z'e\phi$.

Using Eq (3) for c and c', we can rewrite Poisson's equation as

$$d^2\phi/ds^2 = (ne/\varepsilon) \left\{ \begin{array}{l} \alpha z [1 + \exp(-ze\phi/kT) (1 - c_0)/c_0]^{-1} \\ - \beta z' [1 + \exp(z'e\phi/kT) (1 - c_0')/c_0']^{-1} \end{array} \right\} \tag{5}$$

The boundary conditions are

$$\phi = d\phi/ds = 0 \qquad \text{(at infinity)}$$
$$\phi = \phi_0 \qquad \text{(at s = 0)} \qquad (6)$$

The reference point for ϕ in the above boundary condition is different from the one used in Refs. 11-12, but is consistent with Eq (3) and with the convention in the literature on electrostatics.

To determine ϕ_0, we turn to the equilibrium condition for anion vacancy at a perfect source and sink, which is assumed to be the case for the grain boundary. That is,

$$c' = 1/[1 + \exp(z'e\phi_0/kT)\,(1-c_0')/c_0'] = \exp(-F^-/kT) \qquad \text{(at s = 0)} \qquad (7)$$

where F^- is the enthalpy of formation of an anion vacancy. In the extrinsic regime considered here, $1 > c_0' \gg \exp(-F^-/kT)$, ϕ_0 can be approximated very accurately by

$$\phi_0 = (F^-/z'e) + (kT/z'e)\,\ln(c_0'/(1-c_0')) \qquad (8)$$

This potential is always positive because the second term is relatively small in the regime; indeed, ϕ_0 is mostly concentration independent for typical values of F^- (several eV) and kT (0.1 eV). It is this potential drop between s = 0 and infinity that necessitates a charge redistribution in the vicinity of the grain boundary. Also, considering the sign of ϕ, we can verify that the grain boundary is positively charged. Thus, the surrounding space charge, due to the segregation of acceptor solutes, is negative.

The solution of Eq (5), subject to the boundary condition, is straightforward and involves only one numerical integration. This integral is derived in the Appendix and shown below for s > 0 (the functions ϕ, c and c' are symmetric with respect to s)

$$s = \int_\phi^{\phi_0} \left\{ (2nkT/\varepsilon) \left[\begin{array}{l} (\alpha z - \beta z')\,(e\phi/kT) + \alpha \ln\left[c_0 + (1-c_0)\exp(-ze\phi/kT) \right] \\ + \beta \ln\left[c_0' + (1-c_0')\exp(z'e\phi/kT) \right] \end{array} \right] \right\}^{-1/2} d\phi \quad (9)$$

It is easy to verify that at s = 0, $\phi = \phi_0$, and at infinity, $\phi = 0$. Thus, the boundary conditions are indeed satisfied. Having determined ϕ, c and c' can be obtained from Eq (3), the static solution of the electrostatic problem with a stationary grain boundary is then complete.

Some results were obtained by numerical integration of Eq (9) for electrostatic potential for several conditions of c_0, F^-, z and T, at a constant $\varepsilon = 25$ and $n = 3.2 \times 10^{28}/m^3$, which are representative of ZrO_2. Except at very low concentrations, ϕ is almost independent of temperature between 1200 to 1600°C, as is ϕ_0. Its spatial range, reflecting the screening effect, is inversely dependent on the concentration and effective charge. The magnitude of ϕ is also dependent on c_0, z, and F^-, and that of ϕ_0 is most dependent on F⁻ but only weakly dependent on other parameters. The molar fractions of cation solute and anion vacancy are disposed unsymmetrically at the opposite sides of $c = c_0$ and $c' = c_0'$. Almost always c rises to approach the saturation level (unity) in the vicinity of the grain boundary, while c' nearly vanishes there. The ion distributions are nearly temperature independent. The cation saturation is reached more readily at higher effective charge, and when this happens the range increases with the dopant concentration. The range is also inversely dependent on the effective charge. Thus, electrostatic potential and the (negative) space charge (given by $(\alpha zc - \beta z'c')$) are localized within a short distance from the grain boundary due to charge screening by the segregating dopants.

In the following, we examine two limiting cases to obtain a more compact expression of this solution.

Dilute Limit $(c_0 \exp(ze\phi_0/kT)<<1)$: Within a concentration range, this limiting condition can be satisfied while $1 > c_0' >> \exp(-F^-/kT)$ still holds (extrinsic regime). Poisson's equation is reduced in this case to

$$d^2\phi/ds^2 = -(ne/\epsilon) (\alpha zc_0) (z+z') (e\phi/kT) \qquad (10)$$

which has a solution

$$\phi = \phi_0 \exp(-s/\xi) \qquad (11)$$

Here

$$\xi = [(\epsilon kT/ne^2)/(\alpha zc_0) (z+z')]^{1/2} \qquad (12)$$

is the screening distance. Note that ξ^2 varies nearly inversely with the concentration and effective charge of the dopant. This behavior is essentially similar to that found in a dilute electrolyte.[9-10]

Saturation Limit $(c_0 \exp(ze\phi_0/kT)>>1)$: We can simplify the integrand in Eq (9) by noting $e\phi >> kT$ and still letting $c_0 << 1$. Then, the integral becomes

$$s = \int_\phi^{\phi_0} (2n\alpha ze\phi/\epsilon)^{-1/2} d\phi \qquad (13)$$

The solution is

$$\phi = [\phi_0^{1/2} - (n\alpha ze/2\epsilon)^{1/2} s]^2 \qquad (14)$$

To the extent that ϕ_0 is independent of c_0 (which is nearly the case according to Eq (8) when $F^- >> kT$), the solution for ϕ is concentration independent. The screening distance may be defined as

$$\zeta = (2\phi_0 \varepsilon / n\alpha z e)^{1/2} \tag{15}$$

which is inversely dependent on the effective charge of the dopant but independent of concentration. This behavior is new and may be attributed to the saturation of sites in our model. (Within ζ, the solute concentration is essentially unity. The form of Eq (15) can thus be understood by substituting $c = 1$ into Eq (4)).

The above solution is valid for a stationary boundary. In the next section the case of a moving grain boundary is investigated.

INTERACTIONS BETWEEN DIFFUSING SOLUTES AND A MOVING GRAIN BOUNDARY

Diffusion Equation

Returning to the molar fraction of solutes given by Eqs (2-3), we can define the chemical potential of a solute in a field as

$$\mu = U + kT \ln[c/(1-c)] \tag{16}$$

The diffusion flux of solutes in a concentration gradient, subject to a field, follows

$$J = - (Dc/kT) \, d\mu/ds \tag{17}$$

where D is the solute diffusivity. We now let the grain boundary velocity be v. In steady state,

$$J = v \, (c - c_0) \tag{18}$$

where we have used the condition of $c = c_0$ far from the moving grain boundary. Combining Eqs (16-18) gives

$$-dU/ds = kT \{ \, d\ln[c/(1-c)]/ds + v(c-c_0)/Dc \, \} \tag{19}$$

or

$$[D/(1-c)] \, dc/ds + [Dc/kT] \, dU/ds + v(c-c_0) = 0 \tag{20}$$

These equations are slightly different from the ones given in the literature[3-5] in that the concentration dependence has been considered more fully. They can be solved in conjunction with Poisson's equation, Eq (5), subject to the appropriate boundary condition, Eq (6). Such a task, however, is tedious because the potential is now velocity dependent.

Without solving the entire dynamic problem involving a moving boundary, we will establish in the following the necessary relations to evaluate the solute drag from the static solution, derivable from Eq (9).

Force Equation

To determine the frictional force on a moving grain boundary, we can alternatively evaluate the forces required to move solutes. Here we follow the procedure of Hirth and Lothe[11] and develop a force equation and a dissipation theorem which will prove useful shortly. These relations have to be re-established because the diffusion equation in a concentrated solid solution, as given by Eqs (19-20), is different from the one used by Hirth and Lothe for a dilute solution. In addition, the space charge potential U is not merely a classical interaction between a solute and the grain boundary; interactions with other solutes and vacancies are also involved.

We start with the Einstein mobility relation,

$$v = D f / kT \tag{21}$$

where f is the force on an individual solute and v is the velocity of an individual solute. In steady state, this velocity coincides with the velocity of the grain boundary. If we consider only one type of solute for the moment, then all the solutes must experience the same force, Eq (21). The total force, F, is f times $n^* \Delta c$, the excess solutes in total in molar fraction (see below). Here n^* is identified as $n\alpha$ in the case of cation dopants and $n\beta$ in the case of anion vacancies. Therefore,

$$F = (n^*kTv/D)\, \Delta c = (n^*kTv/D) \int_{-\infty}^{\infty} (c-c_0)\, ds \tag{22}$$

where the quantity expressed as the integral on RHS is defined as Δc. Substituting Eq (20) into the above, and recognizing that integration of $(1/(1 - c))dc/ds$ and dU/ds vanishes because of the boundary conditions at infinity ($c = c_0$ and $U = 0$), we can reduce Eq (22) to

$$F = \int_{-\infty}^{\infty} n^* (c-c_0) (-dU/ds)\, ds = \int_{-\infty}^{\infty} n^* c (-dU/ds)\, ds) \tag{23}$$

In hindsight, these results might have appeared self-evident if we had identified $-dU/ds$ as a force on the solute. Such an identification is not justified *a priori* because U is not a classical interaction attributable to the grain boundary itself. Nevertheless, Eq (23) has the same form as Eq (18-36) in Hirth and Lothe for a classical potential.

Although Eq (23) is exact, it is not convenient for evaluating F unless the concentration solution that satisfies Eq (20) for a moving boundary is first obtained. If the static solution of Section 2.2 is used, then by symmetry it leads to $F = 0$ in Eq (23). In the next section, we seek to circumvent this problem by resorting to an alternative expression for F.

Dissipation Theorem

We now proceed to prove a dissipation theorem. The utility of this exercise will become clear shortly. Evaluate the following quantity using Eq (23)

$$Fv = \int_{-\infty}^{\infty} n^* (c-c_0) \, v \, (-dU/ds) \, ds \tag{24}$$

Now using Eq (19) for -dU/ds, and recognizing that integration of $cd\ln[c/(1-c)]ds$ and $d\ln[c/(1-c)]/ds$ vanishes by virtue of the boundary condition, we can reduce Eq (24) to

$$Fv = \int_{-\infty}^{\infty} n^* kTv^2 \, (c-c_0)^2 \, / \, Dc \, ds = (n^* kT/D) \int_{-\infty}^{\infty} J^2/c \, ds \tag{25}$$

Again, Eq (25) has the same form as Eq (18-41) of Hirth and Lothe for a classical potential. An alternative expression of the total force on the grain boundary is thus obtained.

As discussed by Hirth and Lothe, the quadratic form of Eq (25) allows a first approximation of the dissipation, and hence, the total drag force F, to be made by using the static solution for the concentration in evaluating the integral. A comparison of Eqs (22, 25) also reveals the excess solute surrounding the moving grain boundary

$$\Delta c = \int_{-\infty}^{\infty} [(c-c_0)^2/c] \, ds \tag{26}$$

which can be similarly evaluated. Thus, an approximate solution to the solute drag and the excess solute build-up can be obtained without solving the dynamic problem.

MOBILITY OF A MOVING GRAIN BOUNDARY CONTROLLED BY SPACE CHARGE

We are now in a position to evaluate grain boundary mobility which is taken to be controlled by solute drag. Mobility is defined as the ratio of velocity to force. Thus, from Eq (22), and alternatively from Eq (25), we obtain

$$M = v/F = D / n^* kT \, \Delta c$$

$$= (Dv^2/kT) / \int_{-\infty}^{\infty} (J^2/n^* c) \, ds$$

$$= (D/n^* kT) / \int_{-\infty}^{\infty} [(c - c_0)^2/c] \, ds \tag{27}$$

From these expressions, the mobility can be evaluated numerically. Obviously, it is independent of the velocity to the first order approximation.

Of the segregating charged species, one usually diffuses much slower than the rest. Therefore, in almost all cases of interest, we need only evaluate Eq (27) for the species with the slowest diffusivity. In fluorite-structured compounds, this means cations, e.g., Ca^{2+} or Y^{3+} in ZrO_2.[12] In the following, we will examine two limiting cases as in Section 2.2, considering only cation dopants.

Dilute Limit

In this case we let

$$c = c_0 \exp(ze\phi/kT) \tag{28}$$

and, when appropriate, use its Taylor expansion with $ze\phi/kT \ll 1$. Here, ϕ is given by Eq (11). The asymptotic solution of M is found to be

$$M = (D/n\alpha c_0 kT) / \int_{-\infty}^{\infty} (ze\phi/kT)^2 \, ds$$
$$= (D/nkT)(z'^2/z) \, [ne^2(z+z')/\alpha z\epsilon kT]^{1/2} / c_0^{1/2} \, (F^-/kT + \ln c_0)^2 \tag{29}$$

Thus, mobility is nearly proportional to $D/z^{3/2}c_0^{1/2}$. This concentration dependence is weaker than that of the classical case, which predicts M to vary inversely with c_0.[3-5] The difference arises because of the concentration dependence of (a) the screening distance, Eq (12), and (b) the strength of the electrostatic interaction, Eq (8); the latter further tied to the equilibrium molar fraction of vacancies at the source and sink. In general, a slow diffusing dopant of a large effective charge should provide a more effective solute drag.

Saturation Limit

The concentration dependence of mobility in the saturation limit can be evaluated approximately by letting $c = 1$ within a screening distance from the grain boundary and $c = c_0$ outside. Then,

$$M = D / 2n\alpha kT \, \zeta \, (1-c_0)^2 \tag{30}$$

with the screening distance being that of Eq (15). The mobility can be shown to be proportional to $Dz^{1/2}/(1-c_0)^2$. The charge dependence, which reverses the trend in the dilute solution limit, is due to the more effective screening at larger z (ζ decreases with $z^{1/2}$). The concentration dependence is related to the extent of solute segregation, which is limited to $1-c_0$. The above concentration dependence is totally unexpected from the classical theory.[3-5] It predicts that the mobility increases slowly with solute concentration.

We have numerically evaluated the integral in Eq (27) for a number of cases including different effective charge and F⁻. A novel feature, a maximum excess solute or minimum mobility at an intermediate concentration, is revealed. This maximum is not very sharp and the fall-off from it is asymmetric, having a much steeper concentration dependence at lower c_o. The origin of the mobility minimum is clear by comparing the concentration dependence of Eqs (29,30). It corresponds to the maximum of excess solute. These features are nearly independent of temperature, as evident from the previous results on concentrations and electrostatic potentials.

Lastly, for completeness, we state that the contributions of various ions/counterions and intrinsic grain boundary mobility can be accounted for by the following equation

$$M_{total}^{-1} = M_o^{-1} + \sum_i M_i^{-1} \tag{31}$$

where M_o is the intrinsic grain boundary mobility, and M_i^{-1}'s are various contributions due to segregation of ions and counterions evaluated in the same manner as described by Eq (27). The above relation follows directly from the sum of various frictional forces on a moving grain boundary.

DISCUSSION

Several assumptions are central to the present analysis. First, we have assumed a simple lattice model for site occupancy. This leads to a saturation of solute in the vicinity of a grain boundary. Such a model may be oversimplified at high concentrations, and multilayer segregation which follows a much different and more complicated concentration dependence than Eqs (2-3) may need to be incorporated. Second, we have not considered the various solute/defect reactions, including the formation of their complexes. Such reactions do occur in real, ionic solids, and their net effect would be to reduce the concentration of "free" solutes or defects participating in segregation. Thus, the concentration at the mobility minimum is likely to move to a higher value in real ionic solids. Third, we have ignored mechanisms such as polarization energy and compositional dependence of diffusivity and dielectricity. The contribution due to polarization is probably small, judging from the numerical studies on the dipole contribution to space charge by Yan et al.[10] whereas the compositional dependence of diffusivity and dielectric constant will introduce additional spatial variations in the governing Eqs. (5,20) but otherwise not fundamentally alter our description. Fourth, our analysis is limited to the case of low velocity when the static solution may be used in conjunction with Eq (27) to evaluate excess solute and mobility. At higher velocities, the velocity-force relation will most likely become non-linear and a full dynamic solution must be sought. Fifth, we have used fluorite-structured ceramics as a model for analysis to focus attention on cations which are the slower diffusing species. Other ceramics may require a different treatment to account for the possibility of anion diffusion control. Lastly, elastic interactions due to a size misfit or a modulus misfit are not included here as a segregation mechanism since they have been extensively investigated theoretically in the past.[3-5] However, as we pointed out previously, the elastic interactions typically are smaller than the electrostatic potential at least in fluorite structural ceramics.[13]

Our main predictions are now recaptulated below and contrasted with those in the literature. First, regarding space charge distribution around a stationary grain boundary, our result in the dilute limit is, of course, indentical to that found in the conventional electrolyte theory. That is, the screening distance is inversely proportionate to $c_0^{1/2} z^{1/2}$. In the saturation limit, however, we find the screening distance to be inversely dependent on $z^{1/2}$ but nearly independent of c_0. Thus, the double layer cannot be compressed beyond a certain thickness regardless of solute concentration. This limit has never been considered before in the electrolyte literature, possibly because of the lower concentration there and, perhaps more importantly, the much smaller electrostatic potential in liquid electrolytes than in solid state ionics on the effect of F^-. Second, concerning excess solute and boundary mobility, our result in the dilute limit finds the mobility proportional to $D/c_0^{1/2} z^{3/2} (F^-/kT)^2$. This contrasts with the result in the literature (using a classical potential) of $M \propto D/c_0 \xi (z\phi_0/kT)^2$.[3,5] It can be readily verified that the two results are essentially the same except for minor differences between z and z' and $\ln c_0$ and $\ln c'_0$. Lastly, in the saturation limit, we find the mobility reverses its inverse c_0 dependence so that a mobility minimum occurs. This is a totally unexpected result. This mobility minimum is a direct consequence of the segregation model, which incorporates site saturation; otherwise the mobility must decrease monotonically. The higher mobility at a higher z, which also reverses the trend in the dilute solution limit, is due to a more compressed screening distance, which outweighs the effect of an increasing interaction.

The verification of the major predictions of the new solute drag theory has recently come from one family of fluorite-structured solid solutions. In the dilute limit, we used a series of tetragonal zirconia solid solutions[13] and confirmed that the grain boundary mobility decreases with increasing effective charge and decreasing D. A monotonic increase of solute drag with solute concentration was also observed. Some preliminary results in concentrated ceria solid solutions also suggest a mobility minimum as predicted by our analysis. Moreover, the reduction of grain boundary mobility via solute drag has found direct consequence in the deformation characteristics of fine grained zirconia, in that the lowest mobility materials have the lowest flow stress and strain hardening rate due to the stability of microstructure against coarsening[14]. This is perhaps one of the best examples to date of the tailoring of bulk properties through microchemical control of the grain boundaries.

Despite the above encouraging results, we should nevertheless be cautious in applying the solute drag theory blindly. The main reason for this caution lies in the complex effect of solutes on the defect chemistry of ionic solids, rendering diffusion kinetics highly sensitive to solute presence. The effects in most cases are not yet predictable theoretically beforehand and must be established experimentally. Given the dominant role of diffusivity in solute drag, the otherwise clear and predictable charge and concentration dependence of the space charge effect may sometimes be masked.

CONCLUSIONS

Two new theoretical aspects have prompted a more rigorous treatment of the ionic solute drag model. These are (a) the different nature of many-bodied electrostatic interactions which do not have a fixed range; and (b) lattice site competition which leads to site saturation near grain boundaries in an ionic solid of a large electrostatic potential.

This analysis has reaffirmed some known results and revealed surprising new features. As expected, in the dilute concentration limit, the mobility scales with the solute diffusivity and inversely with effective charge and concentration. In the concentrated limit, however, the screening distance is nearly concentration independent, and the mobility goes through a minimum at an intermediate concentration. This minimum recedes to a lower concentration but increases slightly in magnitude as the effective charge increases. Recent experimental results in fluorite-structural oxides are in support of the above predictions.

REFERENCES

1. Yan, M. F., Cannon, R. M. and Bowen, H. K., Grain Boundary Migration in Ceramics. In Ceramic Microstructure '76, eds. R. M. Fulrath and J. A. Pask, Westview Press, Boulder, CO, 1977, pp. 276-307.

2. Aust, K. T. and Rutter, J. W., Trans. Am. Inst. Min. Metall. Pet. Eng., 1959, 215, 119.

3. Cahn, J. W., The Impurity Drag Effect in Grain Boundary Motion, Acta Metall., 1962, 10 [9] 789-98.

4. Lucke, K. and Stuwe, H. P., On the Theory of Impurity Controlled Grain Boundary Motion, Acta Metall., 1971, 19 [10] 1087-99.

5. Srolovitz, D. J., Eykholt, R., Barnett, D. M. and Hirth, J. P., Moving Disconmmensurations Interacting with Diffusing Impurities, Phys. Rev. B, 1987, 35 [12] 6107-21.

6. Hirth, J. P. and Lothe, J., Theory of Dislocations, Second Edition, John Wiley & Sons, New York, 1982, pp. 510-11.

7. Frenkel, J., Kinetic Theory of Liquids, Oxford University Press, New York, 1946, p. 36.

8. Eshelby, J. D., Newey, E., Pyatt, P. and Lidiard, A., Phil. Mag., 1958, 3, 75.

9. Kliewer, K. L. and Koehler, J. S., Space Charge in Ionic Crystals, I. General Approach with Application to NaCl, Phys. Rev., 1965, 140 [4A] 1226-40.

10. Yan, M. F., Cannon, R. M. and Bowen, H. K., Space Charges, Elastic Field and Dipole Contributions to Equilibrium Solute Segregation at Interfaces, J. Appl. Phys., 1983, 54 [2] 764-78.

11. Hirth, J. P. and Lothe, J., Theory of Dislocations, Wiley Interscience, NY, 1982, Chapter 18, and especially pp. 647-9.

12. Oishi, Y., Ando, A. and Sakka, Y., Lattice and Grain Boundary Diffusion Coefficients of Cations in Stabilized Zirconias. In Characterization of Grain Boundaries in Advances in Ceramics, V.7, eds. M. F. Yan and A. H. Heuer, American Ceramic Society, Columbus, OH, 1983, pp. 208-19.

13. Hwang, S. L. and Chen, I-W., Grain Size Control of Tetragonal Zirconia Polycrystals Using the Space Charge Concept, J. Amer. Ceram. Soc., 1990, 73 [11] 3269-77.

14. Chen, I-W. and Xue, L. A., Development of Superplastic Structural Ceramics, J. Amer. Ceram. Soc., 1990, 73 [9] 2585-2609.

APPENDIX—SOLUTION OF POISSON'S EQUATION

Poisson's equation in the form of Eq (5) is formally the same as equation of motion of a particle moving in one dimension in a conservative force field. This analogy is established by the following identification: the coordinate in which the particle travels is ϕ, the "time" coordinate is s, and the mass of the particle is unity. Thus, the LHS of Eq (5) can be regarded as the "acceleration" of a particle, while the RHS is the "force", which is dependent on the position of the particle only. Such a force can be represented as a negative gradient of a "potential energy," $V(\phi)$, as

$$-dV/d\phi = (ne/\varepsilon) \left\{ \begin{array}{l} (\alpha z \left[1 + \exp(-ze\phi/kT)(1-c_o)/c_o \right]^{-1} \\ - \beta z' \left[1 + \exp(z'e\phi/kT)(1-c_o')/c_o' \right]^{-1} \end{array} \right\} \quad (A1)$$

Integration of the above from $\phi = 0$ to ϕ gives

$$V(\phi) = -(nkT/\varepsilon) \left\{ \begin{array}{l} (\alpha z - \beta z')(e\phi/kT) + \alpha \ln \left[c_o + (1-c_o)\exp(-ze\phi/kT) \right] \\ + \beta \ln \left[c_o' + (1-c_o')\exp(z'e\phi/kT) \right] \end{array} \right\} \quad (A2)$$

For motion in a conservative field, the sum of the "kinetic energy" and the "potential energy" is a constant

$$(1/2)(d\phi/ds)^2 + V(\phi) = 0 \qquad (A3)$$

where the constant at RHS is fixed by using the boundary condition at infinity $(d\phi/ds = V(\phi) = 0)$. Therefore,

$$d\phi/ds = - (-2V(\phi))^{1/2} \qquad (s > 0) \qquad (A4)$$

where the negative sign on RHS is fixed by inspection of the boundary condition. Integration of the above, after some rearrangement, gives the solution shown in Eq (9).

Session VI

Interfaces

THE INTERFACE OF MULTIPHASE CERAMIC COMPOSITES

JINGKUN GUO, LITAI MA and BAOSHUN LI

Shanghai Institute of Ceramics, Chinese Academy of Sciences

1295 Dingxi Road, Shanghai 200050, China

ABSTRACT

Studies of particulate dispersion reinforced ZrO_2 composites (PDC), such as Al_2O_3/Y-TZP, SiC(p)/Y-TZP and fibre (or whisker) reinforced ZrO_2 composites, such as SiC(w)/Y-TZP and BN(f)/Y-TZP, have been carried out. The ZrO_2 phase transformation and interface characteristics in different composites with a ZrO_2 matrix have been studied. In all these composites, nucleation is the prerequisite of t-ZrO_2 transformation and the interface characteristics do not seem to have a significant effect except for the existence of interface stress. When an amorphous layer exists at the interface and dissolves some Y_2O_3 from the yttria stabilized ZrO_2 grains, this leads to easier phase transformation of the TZP grains. SiC(w)/Y-TZP or SiC(p)/Y-TZP composites possess better high temperature strength. If the interface reaction layer can be controlled and the strength of the BN fibre improved, the combined effects of fibre reinforcement and phase transformation toughening can be realized in a BN(f)/ZrO_2 composite. Studying the interface can be instrumental in improving the processing of composites.

INTRODUCTION

Tetragonal ZrO_2 polycrystal (TZP) can be fabricated to have very high strength and fracture toughness[1]. However, with increasing temperature, both strength and toughness sharply decrease . Therefore, the study of the phase transformation in tetragonal zirconia (t-ZrO_2) is important. The combined effect of particulate dispersion or fibre (whisker) reinforcement along with phase transformation toughening in ZrO_2 matrix ceramics at high temperatures is also investigated.

The following composite systems: Al_2O_3(p)/Y-TZP, SiC(p)/Y-TZP, SiC(w)/Y-TZP, and BN(f)/Y-TZP have been studied. Special emphasis has been given to microstructural studies and interface characteristics and their consequences on strengthening and toughening.

EXPERIMENTAL

Y-TZP powder containing 2.8 mol.% Y_2O_3 was synthesized by coprecipitating $ZrOCl_2$ + $Y(NO_3)_3$ solution with ammonia and subsequent calcining. The average primary particle size is around 20 nm.

Al$_2$O$_3$ powder was prepared by decomposing ammonium aluminum sulphate. The SiC whiskers were synthesized by the VS method. Boron nitride fibre was provided by the Shangdong Institute for Ceramics Industry. It was processed by nitridation of B$_2$O$_3$ fibre drawn from a melt and then cut to about 10 cm in length. Fine SiC powder (around 5 μm in size) was prepared by ball milling the coarse powder followed by treatment with acid to remove the iron content.

To prepare the composite samples, the fibre (or whisker) was mixed in a high speed mixer with the corresponding matrix powder. After dehydration, the precursor mix was sintered by hot-pressing at appropriate temperatures and pressures. Microstructure observations were carried out under TEM with analytic attachments, and mechanical properties were also determined on a part of the specimens.

RESULTS AND DISCUSSION

1. Al$_2$O$_3$(p)/Y-TZP composite

In the case of Y-TZP ceramics, it has been confirmed that the phase transformation of t-ZrO$_2$ is controlled by nucleation[2,3]. In this study, the interface morphology and intragranular t-ZrO$_2$ transformation of Al$_2$O$_3$(p)/Y-TZP composite were investigated.

Figure 1a is a TEM micrograph of an intragranular t-ZrO$_2$ grain in Al$_2$O$_3$ under constrained conditions. Figure 1b is the dark field of the interface. Figure 2 is a TEM lattice image of the interface of an Al$_2$O$_3$/ZrO$_2$ composite, showing good compatibility between Al$_2$O$_3$ and ZrO$_2$ and little sign of any reaction except for a regular increase in dislocations at the interphase. This condition could aid in reducing stress. Figure 3 demonstrates the process of phase transformation of a ZrO$_2$ grain through nucleation. Figure 3a shows the stress lines existing in a ZrO$_2$ grain. Figure 3b shows the beginning of nucleating transformation and a decreasing stress level in the grain, and Fig.3c illustrates the end of the transformation. Figure 4 shows the sequence of changes of a t-ZrO$_2$ grain free of strain at one edge under strong electron beam irradiation.

Figure 4a is a TEM micrograph of the grain under bright and dark fields. Figure 4b is the electron diffraction pattern of the grain showing its orientation. Figure 4c reveals the process of phase transformation of the grain through nucleation with the help of the irradiation energy although this grain has a size larger than critical. It could be considered as further proof that the prerequisite for t-ZrO$_2$ transformation is the onset of nucleation.

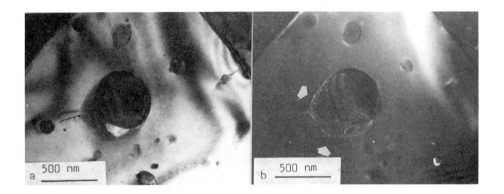

Figure 1 Interface between Al_2O_3 and ZrO_2 (a) and
its image in dark field (b)

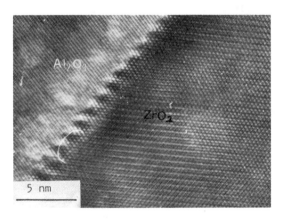

Figure 2 Lattice image of an interface

274

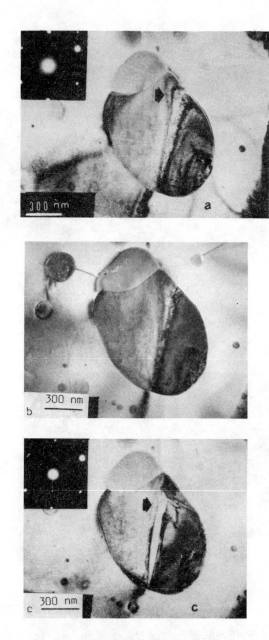

Figure 3 Transformation in Al_2O_3/ZrO_2 composite, showing nucleation and propagation

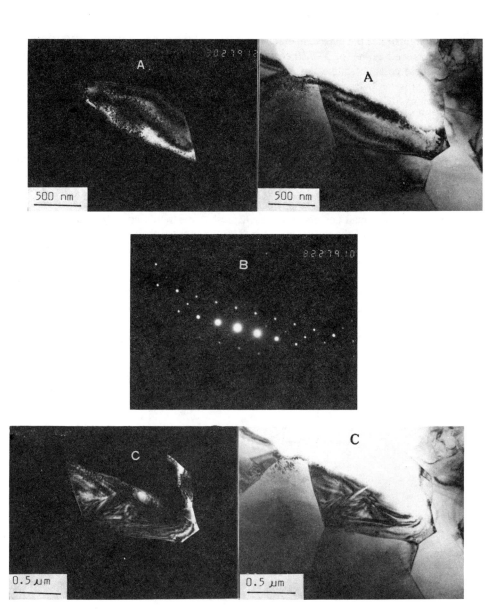

Figure 4 Nucleation and transformation in Y-TZP

a. TEM micrograph of a t-ZrO$_2$ grain with one side free of constraint
b. [011] diffraction pattern of the grain
c. Nucleation and phase transformation

2. SiC(w)/Y-TZP and SiC(p)/Y-TZP composites

Figure 5 illustrates the high temperature strengths of SiC(w)/Y-TZP and SiC(p)/Y-TZP composites as well as that of Y-TZP ceramics, showing that the high temperature strength of both of these ceramic composites is superior to that of Y-TZP, and SiC(p)/Y-TZP is the better one. Figure 6 is a SEM micrograph of SiC(w)/Y-TZP composite, showing the phenomenon of SiC(w) pulled out from the Y-TZP matrix. The room temperature fracture toughness of Y-TZP containing 10 vol.% SiC whisker is 9.3 MPa·m$^{1/2}$ which indicates a considerable toughening effect. Figure 7 is a TEM lattice image of the interface between SiC(w) and Y-TZP, indicating that the interface reaction between SiC and ZrO$_2$ is not excessive and may be helpful for processing and upgrading the properties of this kind of composite. Figure 8 is a TEM micrograph of a SiC/Y-TZP composite, showing the interface glassy phase and the stress lines in the SiC grains. The main composition of the glassy phase is SiO$_2$ formed by the oxidation of the SiC surface. However, EDX analysis has shown that there is a small percentage of Y$_2$O$_3$ present apparently coming from the t-ZrO$_2$ grains. The reduction of the yttria content in ZrO$_2$ grains aids the phase transformation of ZrO$_2$ from tetragonal to monoclinic as shown in Fig.8. Figure 9 shows the microcrystallites in the glassy phase. The microcrystallized glassy phase and microcracks resulting from the phase transformation probably help enhance its mechanical properties at high temperatures. Comparing SiC(p) with SiC(w) composites, SiC particles are more easily and uniformly distributed in Y-TZP than SiC whiskers. This could explain the higher strength of SiC(p)/Y-TZP composites at high temperature than that of SiC(w)/Y-TZP. Therefore, it is important to uniformly distribute the SiC particles or whiskers in the matrix, and if possible, to control the glassy content and increase the content of microcrystallites.

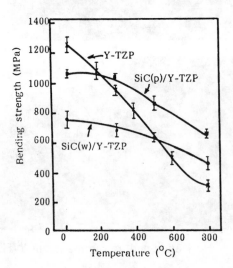

Figure 5 Comparison of the high temperature strengths of
Y-TZP, SiC(w)/Y-TZP and Sic(p)/Y-TZP

Figure 6 A SEM micrograph of the composite, SiC(w)/Y-TZP

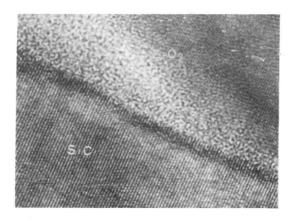

Figure 7 TEM lattice image, showing the interface of
SiC(w)/Y-TZP composite

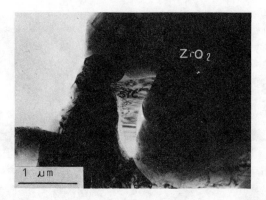

Figure 8   A TEM micrograph of SiC(p)/Y-TZP composite

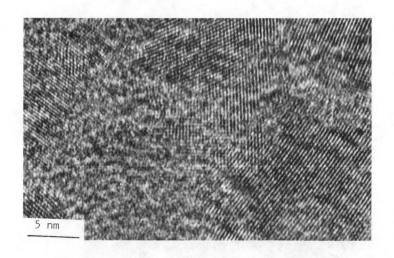

Figure 9   Microcrystallites in the glassy phase
in SiC(p)/Y-TZP composite

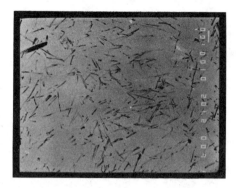

Figure 10 A SEM micrograph of BN(f)/Y-TZP

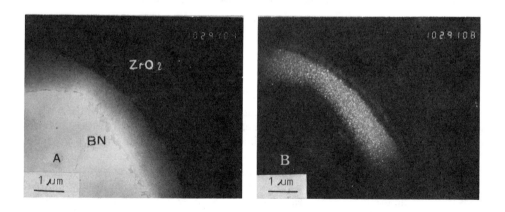

Figure 11 A diffusion region between BN fibre and ZrO$_2$ matrix
in BN(f)/Y-TZP composite (left) and
its dark-field image (right)

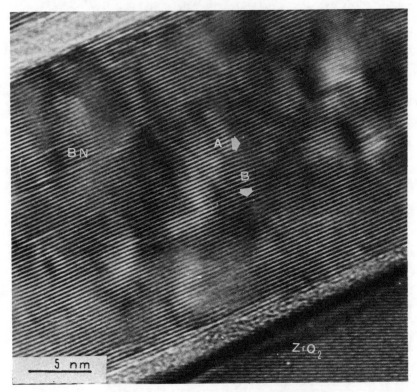

Figure 12 TEM lattice image of BN(f)/Y-TZP composite

3. BN(f)/Y-TZP composite

Figure 10 is a SEM micrograph of BN(f)/Y-TZP composite. The distribution of BN fibre in the matrix does not appear to be uniform. Figure 11 shows a diffusion region with 0.3 μm width at the interface between BN fibre and ZrO_2 matrix. Electron diffraction pattern has indicated the formation of zirconium diboride which is a reaction that can effectively occur at 1600 °C. This reaction zone does not seem to cause t-ZrO_2 transformation. Figure 12 is a TEM lattice image of a BN(f)/Y-TZP composite, showing the interface reaction. It shows the reaction layer at the interface and bright and dark areas in BN structure indicate the existence of stress in the fibre as well as the presence of edge dislocation (A shown in the micrograph) and antiphase domain (B). There is also some glassy phase resulted from the remained and unnitridized B_2O_3 existing in BN fibre as shown in the left part of Fig.11. It can be postulated that to control the purity and defect quality of the BN fibre is essential to enhance the fibre strength. And, in order to get the conbined effect of fibre strengthening and transformation toughening in BN(f)/Y-TZP composite, it has to control the chemical reaction layer between the fibre and matrix.

CONCLUSIONS

1. The prerequisite of t-ZrO$_2$ transformation in either particulate or fibre reinforced Y-TZP composites is the nucleation step.

2. The study of the interfaces could provide useful instructions to improve the technology and performance of the composite materials.

3. As to SiC(p)/Y-TZP or SiC(w)/Y-TZP composite, the interface reaction is not excessive. Both types of composite possess better high temperature strength. It is very important to strictly control the glassy content existing at the interface and if possible make the glassy phase to undergo microcrystallization.

4. Interface reaction exists in BN(f)/Y-TZP composite, but it does not affect t-ZrO$_2$ transformation. By improving the properties of BN fibre and controlling the interface reaction layer, it is possible to achieve a combined effect of fibre reinforcement and transformation toughening.

REFERENCES

1. J.K. Guo and T.S. Yen, "Microstructure and properties of ceramic materials", eds. T.S. Yen and J.A. Pask (China, Beijing, Science Press, 1984) pp.281-290.

2. L.T. Ma, J. Chinese Ceramics Society, 14(2), 169(1986).

3. M. Ruhler, L.T. Ma, W. Wuderlich and A.G. Evans, Physica, 150, 86(1988).

INTERFACES IN SILICON CARBIDE / GRAPHITE LAMINATES

W J CLEGG* and K KENDALL
Solid State Science Group
ICI
PO Box 11 Runcorn Cheshire UK   WA7 4QE

ABSTRACT

Silicon carbide / graphite laminates, made by pressing together 200 μm layers of calendered plastic tape coated with a 3 μm thick graphite film, followed by firing at 2000 ^{O}C, are an order of magnitude tougher than monolithic silicon carbide, while retaining a bending strength of 600 MPa. This extra toughness arises because cracks are deflected along the thin graphite interface layers and so cannot easily penetrate the structure. The background to the deflection of cracks by interfaces is described together with the problems of making tough ceramics by building controlled interfaces into the material. Model experiments illustrate the criteria for interfacial toughening. These criteria are applied to explain the behaviour of the silicon carbide/ graphite laminates.

INTRODUCTION

The use of interfaces to provide crack resistance in brittle materials was analysed by Cook and Gordon [1] in 1964. Their model considered the fracture of a cracked brittle sample (Fig. 1a). When the sample was loaded in tension, the crack moved through the sample catastrophically (Fig.1b). However, if a weak interface was introduced (Fig.1c), the crack was deflected

*now at Department of Materials, Cambridge University, CB2 3QZ

away from its original path (Fig.1d). Fracture was therefore inhibited and the material was tougher, that is more crack resistant.

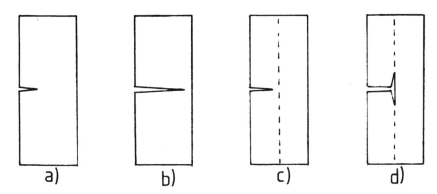

Figure 1. a) Cracked brittle material; b) Catastrophic failure. c) Cracked material with interface; d) Crack deflects.

The example used by Gordon [2] to illustrate this effect was glass reinforced plastic. Both glass and plastic are brittle materials which crack catastrophically. But when combined together, the two brittle materials produce a composite product which is remarkably tough. This toughness can only be explained by the presence of interfaces between the components. The same model could be applied to all brittle materials, ranging from natural organic products like wood, through inorganics like jade and asbestos, to plywood and plastic laminates. Thus, the interfacial toughening concept has led to the wonderful idea that ceramics such as silicon carbide, zirconia, even diamond might be readily toughened by building in the correct interface structures.

This paper outlines some steps on the way to making such tough ceramics economically. First, the problems of building controlled interfaces are recounted. Then the difficulties of theoretical interpretation are analysed using rubber models to demonstrate several important crack deflection effects. Finally, a practical method is described for making tough silicon carbide laminates using graphite as the weak interlayer.

PROBLEMS

There are two problems in achieving this goal of tough ceramics. The first is finding economic methods for building controlled interfaces in ceramic materials. The second is understanding the theoretical mechanisms of interface toughening so that the appropriate composites can be designed.

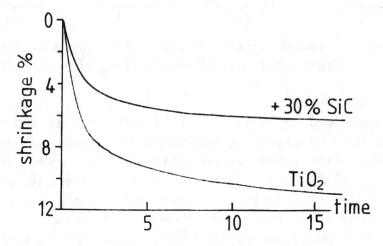

Figure 2 Dilatometer traces showing how the sintering of titania is slowed by additions of rigid particles.

One way of obtaining weak interfaces is to take a ceramic which grows elongated grains (eg Si_3N_4 or SiC) and to dope this with glass which coats the interfaces during sintering [3,4]. If the glass has the correct properties, then cracks do not penetrate the grains but are deflected around them. Another route is to hot press ceramic fibres with glass powder to make glass matrix composites [5-8]. Unfortunately, hot pressing can damage the brittle ceramic fibres, and is expensive. In addition, few strong fibres are available now [9]. A cheaper method is to use ceramic whiskers, mixed with ceramic powder, die pressed and sintered. This method has been used to make SiC reinforced alumina for cutting tools where a modest toughness improvement is valuable [10,11]. Sintering composites in this way, though cheap, is not ideal because sintering is inhibited (Fig.2) by the presence of the rigid fibre [12-14]. Also the rise in properties is small . A better method is to form the ceramic matrix without shrinkage around the fibres by chemical vapour infiltration and deposition. Carbon-carbon and SiC-SiC products are made by this method, which is slow and expensive.

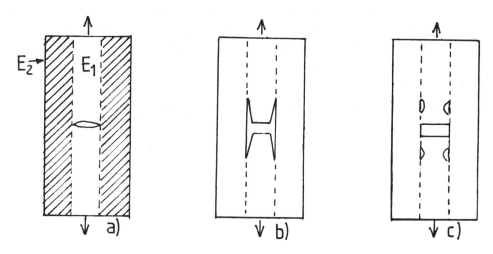

Figure 3. Three mechanisms of interface toughening. a) Crack stopping at a high modulus interface. b) Crack deflection at a brittle interface. c) Formation of interfacial dislocations at a cracked interface.

The second problem is to understand the theoretical mechanisms by which interfaces can toughen ceramics. Cook and Gordon used a stress criterion to define the interfacial condition for crack deflection (Fig.3b). They suggested that the interface should be five times weaker than the bulk material in order to deflect the crack. But this stress criterion cannot be applied to brittle interfaces, as demonstrated by theory and experiment on rubber laminates [15-18]. Fracture mechanics tells us that an energy criterion should be used instead. This result has been confirmed recently by Evans, Hutchinson and colleagues [19-22]. The effect of this debonding is to increase the failure strain of the sample from e_1 to e_2 at the cracking stress σ_1 (Fig. 4b).

Another important crack stopping mechanism was discovered in 1975 [23]. This was the influence of elastic modulus change at an interface (Fig.3a), an effect ignored by Cook and Gordon despite the large difference in modulus between glass and polymer in glass reinforced plastic composites. Cracks cannot easily get out of a low modulus material like plastic to penetrate a high modulus material such as glass. Thus, cracking stress is raised from σ_1 to σ_2 (Fig.4a). By the same token, cracks are attracted across an interface from high modulus into low modulus material, lowering cracking stress.

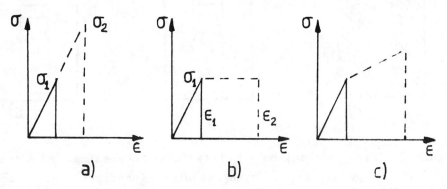

Figure 4. Stress strain curves corresponding to the mechanisms of Fig. 3.

A further mechanism of toughening can occur once the interface has cracked. The crack surfaces are pushed together as the sample is stretched, and adhesion can arise between the surfaces, forming interfacial dislocations, first observed in 1976 [24-26]. An increasing stress is required to drive these dislocations along the interfaces, causing other weak interfaces to be triggered around the crack tip, and creating a wider damage zone (Fig.4c).

THEORY AND MODELS

Fracture mechanics must be used to understand crack propagation along interfaces, based on the Griffith [27] theory that energy supplied to a brittle sample by the applied loads is converted into surface energy of cracks. Using this theory [15] it was shown that the criterion for crack deflection along an interface was

$$R_i/R_c < A \qquad (1)$$

where R_i was the adhesion energy of the interface, R_c was the cohesive fracture energy of the material, and A, the adhesion-cohesion ratio, was a number, typically between .1 and .4 depending on the precise geometry and elastic constants of the materials. Adhesion must be low if cracks are to deflect.

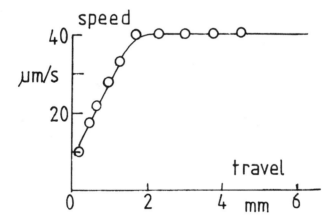

Figure 5. Speed of an interfacial crack with travel after deflection along a weak polymer model interface.

This theory was verified using polymer model experiments, where the cracks could easily be seen and controlled [15-18]. Two blocks of transparent brittle polymer were partly polymerised and then pressed together to make good contact, before completing the polymerisation. This resulted in a sample of uniform elastic modulus, with a plane of deficient polymerisation down the middle, where an interfacial crack could travel. By varying the polymerisation time, the fracture energy of the weak plane could be varied systematically. Cracks were propagated through the block under tension and crack deflection was observed along the interface. The speed of the interfacial crack was found to increase from zero to a steady value at constant load (Fig. 5). By varying the fracture energy at the interface, equation 1 was verified.

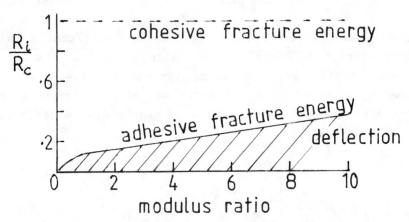

Figure 6. Allowed increase of interface fracture energy to maintain crack deflection as elastic modulus ratio is increased from 1 to 10.

Similar experiments showed that a crack was stopped by a high modulus interface [23]. Theoretically, the strain energy release rate G had to be increased to push the crack through the interface, by an amount given by

$$G_2/G_1 = E_2/E_1 \qquad (2)$$

where E was the Young's modulus of the material and the subscripts referred to each side of the interface. Experiments with polymer samples confirmed this theory. When crack deflection occurred at an interface with modulus mismatch, the adhesion-cohesion ratio A could be increased as shown in Figure 6 [15]. In other words, higher interface adhesion could be tolerated while maintaining crack deflection if the modulus ratio was high. For equal modulus across the interface, the maximum adhesion was 0.09 of the cohesive fracture energy for short cracks, whereas with a modulus ratio of 10, the adhesion could be increased to 0.4 of the cohesive energy.

TOUGH CERAMIC LAMINATES

Both the problems of constructing controlled ceramic interfaces, and of understanding their crack deflection properties, have been addressed in our endeavour to make tough ceramic laminates. Silicon carbide was chosen as the ceramic because of its high elastic modulus together with creep and corrosion resistance. Graphite was selected as the interlayer material because it was known to promote debonding in existing composites, and also because it does not react with SiC even at the highest temperatures [29,30]. The laminar geometry was selected because sheets are easier to handle than fibres. The individual sheets were made from powder for economy, and firing was carried out after lay-up to allow unrestrained sintering.

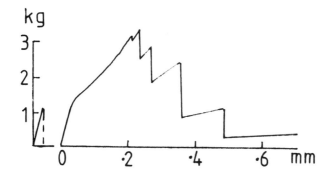

Figure 7. Load/displacement behaviour of a notched SiC/C laminate for comparison with a notched SiC control.

Tapes of silicon carbide powder were made by mixing a 0.2 μm boron doped SiC (Superior Graphite, HSC-059s) with a 40 wt% aqueous solution of polyvinylalcoholacetate (Nippon Gohsei, KH17S) using a high shear mixer. It has been demonstrated that such viscous mixing of fine powders and polymers can cause removal of agglomerates leading to great improvements in strength and reliability of sintered ceramics. The resulting dough-like material was pressed into 2 mm thick sheet and then calendered into flat sheets about 200 μm thick, which were dried and cut into 50x50 mm squares. A graphite slurry was coated onto each face of the squares and the dried, coated sheets were then stacked and pressed into a 2 mm thick slab. After heating this at 1 °C/min in argon to remove the polymer, the samples were sintered at 2040 °C under argon for 30 min. The samples shrank uniformly to give a final gravity of 3.1, almost fully dense.

The composite gave a flex modulus of 450 GPa, strength of 633 MPa and K_{1c} up to 18 MPam$^{\frac{1}{2}}$ measured in a single edge notched bend configuration [35-38]. A typical force/displacement curve for a notched specimen is shown in Figure 7, illustrating the improvement in both load and deflection over the unlaminated control sample, and giving a work of fracture around 6000 Jm^{-2}, about the same as deal wood. By contrast, the work of fracture of sintered SiC was 28 Jm^{-2}, and the fracture energy of the interface was measured as 6.2 Jm^{-2}. Furthermore, the interface adhesion fracture energy could be raised by doping to 16 Jm^{-2} while maintaining crack deflection and toughness.

These numbers seem to go against equation (1) because the adhesion-cohesion ratio measured experimentally was A=0.22 for undoped graphite and 0.57 for doped graphite, far larger than the maximum value of A for crack deflexion of 0.09 calculated theoretically [15]. The reason for this discrepancy is the thickness of the interface layer, which is ten thousand times thicker than atomic dimensions. With very thin graphite layers, it was observed that the crack did not deflect so readily, as predicted by equation (1). For thick graphite, the high value of the adhesion-cohesion ratio can be explained by taking into account the elastic modulus ratio. Here the crack deflection is assisted because the crack cannot easily cross from low modulus graphite into high modulus silicon carbide. This ratio E_2/E_1 amounts to 450/39 or 11.5 for the silicon carbide/graphite interface, as measured on bend test samples of the two separate materials. From Figure 6 it may be seen that the optimum adhesion-cohesion ratio A for crack deflection increases to 0.45 at this condition, much closer to the experimentally observed values.

CONCLUSIONS

Cracks in ceramics can be deflected by brittle interfaces, giving a substantial increase in fracture toughness. The two basic criteria describing this effect; the adhesion-cohesion ratio and the elastic modulus ratio; were defined in the early 70s [15,23] from fracture mechanics principles and have since been further elucidated [28]. These criteria allow tough ceramic laminates to be designed. There is the further problem of building interfaces economically, since hot pressing and vapour deposition are expensive. This problem has been overcome by making laminates from calendered plastic sheets made from sub-micron ceramic powders. Thin sheets are coated with a slurry of interface agent, then pressed into laminates which are pressureless sintered. Silicon carbide/graphite laminates made by this method have been shown to give ease of fabrication, high toughness, stiffness and strength [38].

REFERENCES

1. Cook,J and Gordon, J.E., Proc R Soc Lond,1964, A282,508.
2. Gordon, J.E., The new science of strong materials, Penguin, London 1968 pp. 106-8.
3. Neil, J.T., Pasto, A.E. and Bowen, L.J., Adv Ceram Mater,

1988, 3 225-230.
4. Boecker, W.G.D.,Storm, R.S. and Chia, K.Y.,EP419271142 1990.
5. Philips, D.C., J Mater Sci, 1974, 9, 1847-1854.
6. Philips, D.C., Sambell, R.A.J. and Bowen, D.H.,J Mater Sci, 1972, 7, 1454-64.
7. Prewo, K.M. and Brennan, J.J., J Mater Sci, 1980, 15, 463-8.
8. Prewo, K.M., Bull Am Ceram Soc, 1989, 68, 395-400.
9. Yajima, S., Hayashi, J., Omori, M. and Okamura, K., Nature, 1976, 261, 683-4.
10. Wei, G.C. and Becher, P.F.,Bull Am Ceram Soc, 1985, 64,298-304.
11. Baldoni, J.G. and Buljan, S.T., Bull Am Ceram Soc, 1988, 67, 381-87.
12. De Jonghe, L.C., Rahman, M.N. and Hsueh, C.H.,Acta Met, 1986, 34, 1467-71.
13. Clegg, J.W., Alford, N.McN., and Birchall, J.D., Proc Brit Ceram Soc, 1987, 39, 247-254.
14. Clegg, W.J. and Birchall, J.D., Proc 4th Int Conf FRC, Inst Mech Eng, London, 1990, p.179.
15. Kendall, K., Proc R Soc Lond, 1975,A344,287-302.
16. Kendall, K., J Mater Sci, 1976, 11, 638-644.
17. Kendall, K., J Phys D: Appl Phys, 1975, 8, 512-522.
18. Kendall, K., Mat Res Soc Symp Proc, 1985, 40, 167-176.
19. Evans,A.G., He, M-Y. and Hutchinson, J.W., J Am Ceram Soc, 1989, 7, 2300-2303.
20. Charalambides, P.G., Lund,J., McMeeking,R.M. and Evans, A.G., J Appl Mech 1989, 56, 77-82.
21. He, M-Y. and Hutchinson, J.W., J Appl Mech, 1989,56,270-78.
22. Evans, A.G., Ruhle, M., Dalgleish, B.J. and Charalambides, P.G., Mater Sci Engng, 1990, A126, 53-64.
23. Kendall, K., Proc R Soc Lond, 1975, A341, 409-428.
24. Kendall, K., Nature, 1976, 261, 35-6.
25. Kendall, K., Phil Mag, 1977, 36, 507-15.
26. Kendall, K., Phil Mag, 1981, 43, 713-29.
27. Griffith, A.A., Phil Trans R Soc Lond, 1920, A221, 163.
28. He,M-Y, and Hutchinson, J.W., Int J Solids Structures, 1989, 25, 1053-1067.
29. Corbin, N.D., Rossetti, G.A. and Hartline, S.D., Ceram Engng Sci Proc, 1985, 6, 632-645.
30. Birnie, D.P., Mackrodt,W.C. and Kingery,W.D., Adv Ceram, 1986, 23, 571-584.
31. Alford, N.McN., Birchall, J.D. and Kendall, K., Nature, 1987, 330, 51-53.
32. Kendall, K., Powder Met, 1988, 31, 28-31.
33. Kendall, K., Materials Forum, 1988, 11, 61-70.
34. Kendall,K.,Alford,N.McN.,Clegg, W.J. and Birchall, J.D., Brit Ceram Proc, 1990, 45, 79-89.
35. Clegg. W.J., Kendall, K., Alford, N.McN., Button, T.W. and Birchall, J.D., Nature, 1990, 347, 455-7.
36. Clegg, W.J. and Seddon, L.R., Proc 4th Int Symp Ceram Mater for Engines, Goteborg, Sweden 1991 in press.
37. Clegg,W.J. and Seddon, L.R., Proc 2nd Euro Conf Adv Mater and Proc, Cambridge UK, 1991, in press.
38. Clegg, W.J. and Kendall, K., EuroPatent GB9002986 1990.

HYBRID MATRIX COMPOSITES REINFORCED WITH CARBON FIBRES VIA POLYMER PYROLYSIS

M. IWATA, A. NAKAHIRA*, H. INADA and K. NIIHARA*
Noritake Co.,Ltd.* ISIR, Osaka University.
300 Higashiyama, Miyoshi-cho, Nishikamogun, Aichi.
*8-1 Mihogaoka, Ibaraki 567, Osaka.

ABSTRACT

Polymer pyrolysis, combined with filament winding, was adopted for manufacturing carbon fibre reinforced Si-Al-O-N matrix composites with dispersed SiC particles.

Flexural strength, fracture toughness, and debonding strength were significantly improved by incorporating very fine β-SiC particles, as expected from the already known nanocomposite effect.

The particular finding is that the high stiffness of the Si-Al-O-N matrix composite was maintained up to 1300°C as a result of the presence of dispersed SiC particles.

INTRODUCTION

Much attention has been focused on improving the brittleness of ceramic materials by continuous fibre reinforcement. This is one of the most promising methods to toughen ceramic materials. Several processing methods such as, slurry infiltration (1-2), sol-gel (3), polymer pyrolysis (4-5), and in-situ chemical reaction (6-9) techniques, have been employed for fabricating continuous fibre reinforced ceramic composites.

Polymer pyrolysis, followed by sintering using either a gas-pressure furnace or a hot-press furnace, is the best technique to form ceramic materials. This process is closely based on the structures of organometallic polymers as the precursors to ceramics.

But the problem with this process is that very high sintering temperatures are required to convert organometallic polymers into ceramic matrices. Thus, the usual ceramic fibres such as Nicalon® and Tyranno® fibres can not be selected as the reinforcement. The possible fibre candidate for this process is a carbon fibre, which gives good mechanical properties and is compatible up to very high temperatures.

On the other hand, Niihara and coworkers recently succeeded in using standard sintering techniques to fabricate ceramic nanocomposites in which the nanosize particles were dispersed in the matrix grains. These nanocomposite techniques were found to improve the mechanical and chemical properties of Al_2O_3, MgO, and Si_3N_4 among others, even at high temperatures.

The objective of the present work is to prepare hybrid matrix composites. Specifically, to evaluate the mechanical properties at very high temperatures of Si-Al-O-N matrix composites containing a very fine dispersion of β-SiC particles and reinforced by carbon fibres, fabricated by polymer pyrolysis combined with filament winding.

MATERIALS AND METHOD

Raw materials

The high modulus carbon fibres used in this study are HMS-55X fibres produced by Toho Rayon Co.,Ltd. These fibres, 12,000 per strand, are composed of a crystalline graphite phase and an amorphous phase. Their average tensile strength and elastic modulus are 3970 MPa and 555 GPa, respectively.

Polysilazane, NCP-201, produced by Chisso Corporation, has an average molecular weight of 1050. Fine α-Si_3N_4 (E-10 powders, produced by Ube Industries Ltd.) and γ-Al_2O_3, (AL-20 powders, produced by Asahi Chemical Industry Co.,Ltd.) were selected as reactive fillers. The average diameter of the α-Si_3N_4 and γ-Al_2O_3 were 0.1 μm and 0.4 μm, respectively.

Fine β-SiC powders produced by Ibiden Co.,Ltd., with a diameter of 0.25 μm, were used as the second phase dispersoids.

Fabrication of Ceramic Composites

Unidirectional fibre aligned preform tape was made by winding slurry-infiltrated strands onto a mandrel (100mm in diameter) at a strand-supplying-speed of 5 cm/sec, as illustrated in Figure 1. The slurry consisted of a toluene solvent, polysilazane, α-Si_3N_4 and γ-Al_2O_3 with a 10 mol% dispersion of β-SiC, which was mixed before preparing the slurry. The preform tape was cut into 90 mm-long, 50 mm-wide segments to fabricate the postforms.

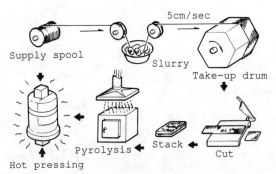

Figure 1. Schematic fabrication process for fibre reinforced ceramic composites.

The postforms, encapsulated into thin polymer films such as polyethylene films in a vacuum, were then isostatically pressed (stacked), dried completely for 48h between 70°C and 80°C, and pyrolyzed at 550°C under a N_2 pressure of 0.5 MPa.

They were then hot-pressed in an Ar atmosphere at 1600°C, 1650°C, 1700°C, and 1750°C under an applied pressure of 33 MPa, to convert them into the composite products.

Characterization of Composites

The characteristics of the sintered composites were measured with respect to bulk density, apparent porosity, and water absorption. Crystal phase identification was done by X-ray diffraction analysis.

Young's modulus, flexural strength, and fracture toughness were evaluated using specimens of 3mm x 4mm x 40mm. Young's modulus was estimated using the flexural vibration method. The flexural strength tests were done using the 3-point bending method with a span of 30 mm at a crosshead speed of 0.5 mm/min both at room temperature, and at 1000°C, 1200°C, 1300°C, 1400°C, and 1500°C in an Ar atmosphere. The fracture toughness tests at room temperature were performed via the SEVNB method (10) with a span of 30 mm at a crosshead speed of 0.5 mm/min using specimens having a V shaped notch. Fracture toughness was calculated using equations (1) and (2).

$$K_{IC} = (P_{max} L/BW^{3/2}) [(3/2) (a/w)^{1/2} Y (a/w)] \qquad (1)$$

$$Y (a/w) = 1.964 - 2.837 (a/w) + 13.711 (a/w)^2$$
$$-23.250 (a/w)^3 + 12.129 (a/w)^4 \qquad (2)$$

where the maximum fracture load is P_{max}, the specimen thickness is B, the specimen width is W, the notch depth is a, the span of the 3-point bending tests is L, and the shape factor is Y.

RESULTS AND DISCUSSION

Characterization of Composites

Table 1 lists the crystalline phases in the ceramic composites hot-pressed at 1600°C, 1650°C, 1700°C and 1750°C. The composite, hot-pressed at 1600°C, contained untransformed α-Si_3N_4, Si-Al-O-N ($Si_3Al_{2.67}O_4N_4$), X-sialon, and X'-sialon, which were synthesized by the in-situ reaction between Si_3N_4 derived from polysilazane, γ-Al_2O_3, and α-Si_3N_4. For the composites hot-pressed at 1600°C, 1650°C, 1700°C and 1750°C, the main crystalline phases were Si-Al-O-N and X or X'-sialon, and the secondary crystalline phases, α- or β-Si_3N_4 were only slightly detected. None of the SiC peaks were detected by X-ray analysis because the estimated content of SiC dispersed in the matrix was only 1.3 wt%.

However, transmission electron microscopic observation revealed that β-SiC particles smaller than approximately 0.1 μm were mainly located within the Si-Al-O-N matrix grains.

TABLE 1

Crystalline phases in ceramic composites by X-ray diffraction analysis, excluding crystalline graphite phases

Sintering temperature (°C)	Crystal phase
1600	$Si_3Al_{2.67}O_4N_4$, α-Si_3N_4, X
1650	$Si_3Al_{2.67}O_4N_4$, X , β-Si_3N_4
1700	$Si_3Al_{2.67}O_4N_4$, X or X', β-Si_3N_4
1750	$Si_3Al_{2.67}O_4N_4$, X , β-Si_3N_4
	X:$Si_6Al_6O_9N_8$, X':$Si_{12}Al_{18}O_{39}N_8$

Bulk densities, appartent porosities and water absorptions of the composites are shown in Table 2.

TABLE 2
Sintering characteristics of ceramic composites reinforced by carbon fibres

	Sintering temperature (°C)			
	1600	1650	1700	1750
Bulk density (g/cm³)	2.42	2.48	2.57	2.52
Apparent porosity (%)	7.63	5.07	1.72	2.99
Water absorption (%)	3.16	2.08	0.67	1.19

As can be seen in Table 2, sintering behavior was strongly influenced by the sintering temperature. The best properties were obtained by hot-pressing at 1700°C, at which a maximum bulk density of 2.57 g/cm³, a minimum apparent porosity of 1.27%, and a minimum water absorption of 0.67% were obtained.

Mechanical Properties of Composites

Figure 2 shows the variation of the flexural strength with the sintering temperatures for the Si-Al-O-N/SiC nanocomposites reinforced with carbon fibres. A flexural strength of 314 MPa was observed for the Si-Al-O-N matrix composite without the SiC particle dispersion. As seen in Figure 2, however, the flexural strength was tremendously improved by incorporating very fine β-SiC particles in the Si-Al-O-N matrix grains. The maximum flexural strength, 705 MPa, was obtained by hot-pressing at 1700°C.

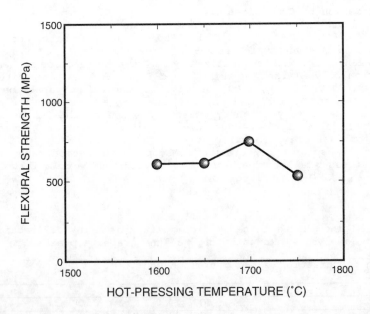

Figure 2. Relation between flexural strength and hot-pressing temperature.

The fracture toughness was also markedly improved by the very fine dispersion of β-SiC particles in the interior of the Si-Al-O-N matrix grains, as shown in Figure 3. The fracture toughness values were 12.3 MPa·m$^{1/2}$ at a sintering temperature of 1600°C, 20.6 MPa·m$^{1/2}$ at 1650°C, 23.5 MPa·m$^{1/2}$ at 1700°C, and 20.2 MPa·m$^{1/2}$ at 1750°C.

By comparison, a fracture toughness of 9.8 MPa·m$^{1/2}$ was observed for the Si-Al-O-N matrix composite without the SiC particle dispersion, prepared using the same sintering technique under similar conditions. In Figure 4, the typical load-displacement curve of the composite with the SiC particle dispersion was compared with that of the Si-Al-O-N matrix composite without the SiC particle dispersion.

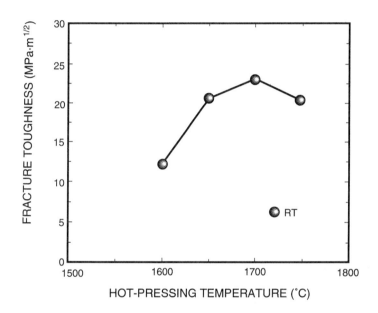

Figure 3. Comparsion of fracture toughness at room temperature, related to hot-pressing temperature.

This comparison emphasizes the marked advantage of a very fine dispersion of β-SiC particles in the Si-Al-O-N matrix grains. The load for debonding between the carbon fibres and the matrix is high enough to maintain the maximum load, compared with the Si-Al-O-N matrix compsite without the SiC particle dispersion, in which the load for debonding is approximately 60% of the maximum load. The debonding stress calculated using the load for debonding that appears in the load-displacement curve was maintained up to 83% to 92% of the maximum stress. The smallest difference between the fracture stress and debonding stress was obtained at a sintering temperature of 1700°C. Regardless of the small difference, both the bridging and pull-out behavior, which are very important phenomena for improving fracture toughness by fibre reinforcement, were observed after debonding between the matrix and the carbon fibres occurred.

From these results, a sintering temperature of 1700°C is the nearly optimum condition to control chemical as well as physical interactions between the carbon fibres and the matrix. The reason that the interfacial structure between the Si-Al-O-N matrix and carbon fibres is optimized by the presence of a very fine β-SiC particle dispersion within the Si-Al-O-N matrix grains is not clear at present. Further studies are required.

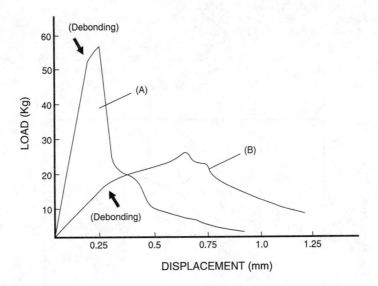

Figure 4. Load-displacement curve of the Si-Al-O-N matrix composites both with (A) and without (B) the SiC particle dispersion.

The Young's modulus of the composite increased gradually up to 262 GPa with increasing sintering temperatures. This value is approximately 66% of the average Young's modulus of 400 GPa, calculated using the composite rule. Where the carbon fibre content is 50 vol%, the Young's modulus of the carbon fibre is 555 GPa and the Young's modulus of the matrix is 250 GPa. The Si-Al-O-N matrix composite containing a SiC particle dispersion retained its stiffness up to 1300°C. This suggests that the Si-Al-O-N matrix maintains its structural integrity up to at least 1300°C. The flexural strength at 1300°C was approximately 1.5 times as high as it was at room temperature. Based on the effects of the SiC particle dispersions in the Al_2O_3, MgO, and Si_3N_4, based nanocomposites, it is reasonable to suggest that these improvements are achieved due to the very fine β-SiC particles dispersed both intragranurally within the grains of the Si-Al-O-N matrix and intergranurally at the grain boundaries of the Si-Al-O-N matrix. (11-13)

CONCLUSION

Carbon fibre reinforced Si-Al-O-N matrix composites with dispersions of nanosize SiC particles were fabricated using polymer pyrolysis and hot-pressing combined with the filament winding.

Flexural strength, fracture toughness, and debonding strength were significantly improved by dispersing very fine β-SiC particles both at the grain boundaries of the Si-Al-O-N matrix and within the grains of the Si-Al-O-N matrix.

The Si-Al-O-N matrix composite containing a SiC particle dispersion showed a 3 times higher debonding strength (646 MPa) and a 1.5 times higher flexural strength (1078 MPa) at 1300°C.

REFERENCES

1. Prewo, K.M. and Brennan, J.J., J.Mater.Sci., 1989, 15 (2), 463-468.
2. Prewo, K.M. and Brennan, J.J., J.Mater.Sci., 1982, 17 (4), 1201-1206.
3. Fitzer, E. and Gadow, R., Conf. on Tailoring Multiphase and Composite and Composite Ceramics, Penn. State Univ., 1985.
4. Oshima, K., Iwata, M. and Isoda, T., Proc. of the 2nd Japan Int. SAMPE symp., 1991, PP.836-843.
5. Iwata, M., Nakahira, A., Inada, H., and Niihara, K., Proc. of the 1st Int. Symp. on the Sci. of Eng. Ceram., 1991, PP.387-391.
6. Caputo, A.J. et al., Ceram. Eng. Sci. Proc., 1985, 6-7/8, 694-706.
7. Caputo, A.J. et al., Ceram. Eng. Sci. Proc., 1984, 5, 657-667.
8. Kennedy, K. and Newkirk, M.S., U.S.Pat.4, 824, 622, 1990.
9. Kesher, H.D. and Kennedy, C.R. et al., U.S.Pat.4, 921, 818, 1990.
10. Awaji, H., Sakaida, Y. and Watanabe, T., Proc. of 6th Int. Conf. of Mechanical Behaviors of Materials-VI (ICM VI), 1991, PP.503-508.
11. Niihara, K., Nakahira, A., Sasaki, G., and Hirabayashi, M., Proc. MRS Int. Meeting on Advanced Materials., 1989, 5, 129-134.
12. Niihara, K., Hirano, T., and Nakahira, A., et al., Proc. 1st Int. SAMPE Symp., 1989, PP.1120-1125.
13. Niihara, K., Hirano, T., Nakahira, A., Proc. of MRS Int. Meeting on Advanced Materials., 1989, PP.107-112.

STUDY OF SOL-GEL TRANSITION IN CERAMIC SYSTEMS BY HIGH RESOLUTION REFRACTOMETRY.

RICARDO CASTELL, ANA RITA DI GIAMPAOLO, HERVE PERRIN, CARMEN
SAINZ*, ANTONIO GUERRERO, JOSE CALATRONI and JOAQUIN LIRA
National Research Group on Surfaces and Interfaces
Engineering.Laboratory of Optics and Electron Microscopy. Simon Bolivar
University. P.O.Box 89000, Caracas 1080 A, VENEZUELA.
* Metropolitan University, Caracas, VENEZUELA.

ABSTRACT

Sol-gel methods are used to synthesize ceramic oxides for a wide range
of applications in coatings, powders and monolithic pieces. The early
stages of sol reactions, up to gelation, play an important role in the
nanoscale microstructure , morphology and density of gels. This paper
presents a recently developed interferometric method. A spectroscopic
analysis of interferograms of white light passing through samples of
sols, permits the precise measurement of gelation time up to 10^{-7} s,
and the geometrical positioning of the initial stage of the gel in the
solution.

INTRODUCTION

The sol-gel method is a widely used to synthesize ceramic oxides.
It has a variety of applications because the products can be formed at
low temperature into fibers, coatings, powders and monolithic parts (1).
It is well recognized that the earlier stages of sol reactions, until sol-
gel transition, play an important role in the nanoscale structure of gels
as well as the morphology and density at different length scales of the
ceramic products. Nuclear magnetic resonance (NMR), Raman, and
infrared spectroscopy, small angle scattering of X-rays (SAXS) and
viscosity measurements have all been used for determining the structure
of sols and gels and the gelation time. A review of these methods has
been published recently by Brinker and Scherer(2).

The refractive index of a transparent medium, is an important
parameter owing to its correlation with other physical variables of the

sample. Temperature, concentration, density, viscosity, and residual stresses, are frequently measured using the refractive index, where the optical path of the beam in the device is modified in the presence of the sample(3). Among the many optical methods available to observe modifications of the optical path , interferometric processes have proven to be the most precise. Furthermore, as these are basically differential procedures they are particularly applicable to those problems where relative changes of the refractive index, $\Delta\eta$ (η_{sample}-$\eta_{reference}$), are anticipated.

Recently(4), a new interferometric process for the measurement of the refractive index has been introduced. It involves the spectroscopic analysis of an interferogram formed by a beam of white light, that is, by continuous spectral distribution. This procedure has been called, White Light Interferometry with Spectral Resolution, allows precision measurements of differences in the index of refraction $\Delta\eta$, up to 10^{-7} at any point in the sample.

This research presents an application of interferential refractometry to sol- gel transitions in a SiO_2 system

Basic Principles

The principles of white light interferometry using spectral resolution have been presented recently elsewhere(5). High resolution spectral interferometry, involves analyzing the spectrum of a white light source. Figure 1 shows a schematic diagram of the equipment consisting of a Michelson interferometer, and a spectrometer. The specimen and the reference sample, are mounted in transparent containers at each side arm of the interferometer. The continuous spectrum of white light produces sets of monochromatic fringes at the exit plane, a set for each wave length present in the spectrum. All these sets of fringes are superposed at the (x', y') exit plane. The spectrometer's slit selects a line (y=0) on this plane and receives light only from this vertical line on the specimen. Thus the spectrometer splits up the superposition of monochromatic fringes present at each point of the slit. This yields the intensity of the integrated fringe system as a function of each monochromatic component (i.e. λ). If now a two dimensional detector, like a flat arrangement of photodiodes, is placed at the plane (x", λ), a spectral distribution can be registered along a line on the sample.

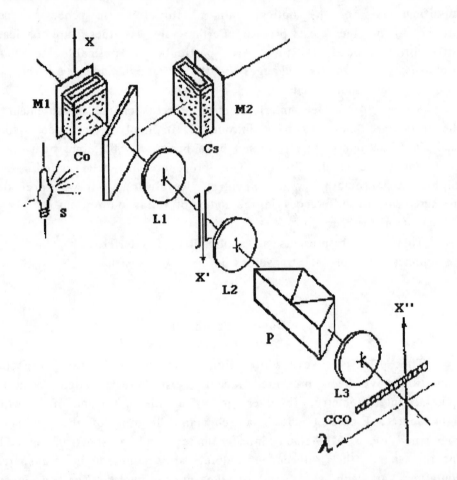

Figure 1: Schematic of experimental set-up. S is the light source. M1 and M2 are the mirrors of the interferometer. C_e and C_s are the reference and the sample cells (or holders), and P is the prism (dispersive element) of the spectrometer. The x"-λ plane is the exit plane where the Charge Coupled Device Television Camera (CCD-TV) is placed.

Figure 2 shows a characteristic spectral distribution for a point on the sample. The intensity oscillations indicate its harmonic variation with wave length. The period of these oscillations is inversely proportional to the path difference in the interferometer. The spectral distribution of the light intensity is correlated to the phase difference between the spectrometer's arms. At the same time, the phase differences are a linear function of $\sigma=\lambda^{-1}$ and the difference between the refractive indices of the specimen and the reference sample.

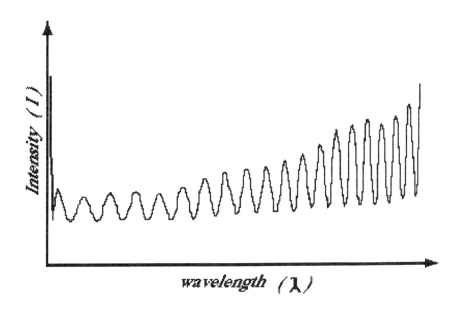

Figure 2. A characteristic spectral distribution for a point of the sample

For small variations of the refractive index along the specimen (as is usually the case when a precision interferometric measurement is needed) the phase difference is a function of σ. Thus the graphic representation of phase differences as a function of σ is a straight line with a slope $e\Delta\eta$, where e is the thickness of the specimen's transparent container. This corresponds to the case where the dependence of $\Delta\eta$ on $\Delta\sigma$ can be disregarded in the spectral interval covered by the spectrometer. This is a very important physical situation because it allows the determination of the slope of the phase differences as a function of σ with high precision. By measuring this slope for each point on the sample, $\Delta\eta$ can be calculated. This is the only reason for working with white light.When $\Delta\eta$ is independent of σ the spectral analysis of the interferogram using white light makes it possible to obtain as many independent measurements of $\Delta\eta$, as there are different wavelengths in the spectral band's width from the light source. For a detector built from a flat arrangement of photodiodes (Charge Coupled Device-CCD), the number of wavelengths resolved for the system is of the order of 10^3. Thus an important improvement in precision is obtained, compared with interferometric measurements using a single wave length, if each measurement is the result of approximately 10^3 independent measurements of $\Delta\eta$, one for each wavelength present in the spectrum.

EXPERIMENTAL

Sol preparation was carried out by hydrolysis and condensation of tetraethylorthosilicate (TEOS). A transparent solution was obtained by mixing TEOS with ethanol, HCl, and H_2O in the molar ratio 1:4, 1: 1.5 and 1:6, respectively. The solution was immediately poured into the sample holder. The specimen and the reference sample were then introduced into the transparent sample holders, The distilled water reference is introduced only to avoid dispersion effects (see Fig1). The light source was a 200 W Xe lamp. A classic spectrometer, operating in the 5500 Å-6500 Å range, was used. A CCD video camera with a flat arrangement of 512 x 512 photodiodes, was positioned at the exit plane. The spectral resolution with this arrangement is of approximately 3Å, The analog signal of the video camera is quantified in 256 levels of gray, and sent to a microcomputer. The recording time for a full image is 0.3 s. From the record of the fringe system at the x", λ plane, using appropiate software program, the optical phase as a function of σ for each point of the sample can be obtained and from it the value of $\Delta\eta$ for each point of the sample. In order to correct for apparent phase difference variations due to variations of the thickness of the specimen holders, an initial reference run is made. The phase variations obtained during gelation are subtracted from the the reference image.

A first experiment to determine $\Delta\eta$ was run in order to ascertain if the index variations in the sol-gel process were within the range of interferential refractometry. A schematic of the experimental set-up is shown in Figure 3. The sample solution is placed in the triangular base of hollowed prism and illuminated by a He-Ne laser. The beam deflection is registered as a function of time thereby obtaining the evolution of the refractive index with time. Water is used in the second prism, because its refractive index is relatively close to that of the sample (η_{sample} =1.369).

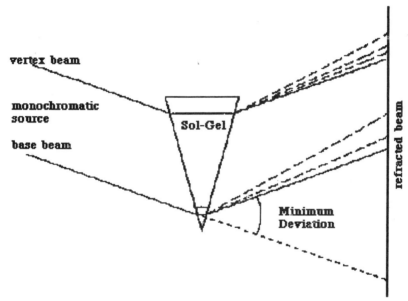

Figure 3. Schematic of the experimental set-up for the classical observation of the evolution of the refractive index with time. The sol-gel sample solution is contained in the triangular base of a prismatic sample holder.

A second set of experiments was run using parallel walled prisms and a white light source. This permits the analysis of the interference lines over a wide range of wavelengths, thereby allowing the investigation of the effects of sol thickness on the formation time of gel primary particles, surface effects and the accurate determination of gelation time.

Results

Figure 4, shows the variation of the refractive index is plotted as a function of time for two different paths of the light beam within the prism: one close to the vertex and the other close to the base. It can be seen that the gelation process, interpreted as an increment of the refractive index, is faster in the narrower zone (near the vertex) than in the wider zone. This experiment is the first quantitative demonstration of the dependence of the sol-gel transition on the thickness of the container. The sol-gel transformation was completed in 450 min with a refractive index increment of $\Delta\eta=5\times10^{-3}$. Based on this value, the interferometric assembly was mounted, Figure 2 shows a typical fringe structure which, when processed, yields a $\Delta\eta$ value with a precision of 10^{-6}. These interferometric experiments confirm the evolution shown in Figure 4.

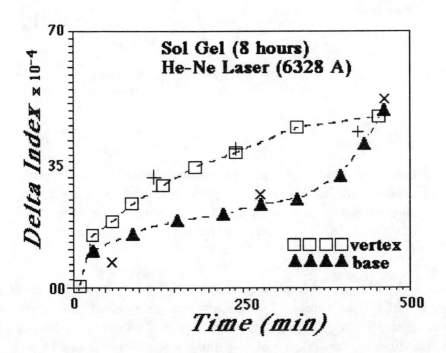

Figure 4. Graphic representation of the refractive index changes with time Note the variation with thickness of the sample (see Fig.3).

The interferometric analysis of the solution placed in the parallel walled prism (Fig 5) using white light, yielded refractive index changes with time along a vertical line of the sample that increased with time up to 250 min (~ 4 h).

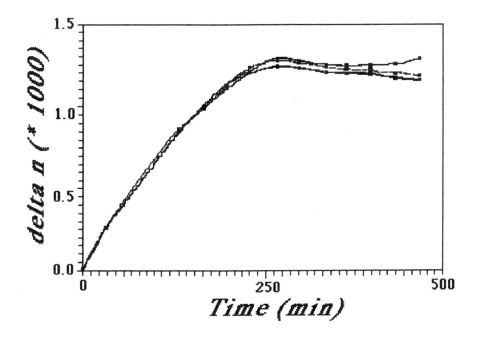

Figure 5: Spectrally resolved white light interferometry. The curves show the changes of the index of diffraction with time, at three different points along a vertical axis on the sample, at 1 mm intervals

No appreciable differences in the slope when the index of refraction is measured at various sample depths in the container can be observed during the constant increase, however after 4 h, $\Delta\eta$ starts to decrease slowly, indicating the culmination of the gelation process and the on set of densification of the gel. Measurements taken at different heights along a vertical line on the sample at this new stage show rate differences for $\Delta\eta$. This indicates the influence of the free surface of the sample on the refractive index, thus on gelation, or primary particle formation. This could be due to rapid evaporation of the volatile alcohols from the sample.Using this method very accurate time for gelation can be measured as well as measureing accuratly the refractive index. Experimental errors are about 10^{-7} compared with 10^{-3} with geometrical methods and 10^{-4} to 10^{-6} with classic interferometric systems. A possible explanation for the negative slope of the curve might be as a consequence of thermal effects.

CONCLUSIONS

The present study has shown that interferometric patterns analyzed by means of diffractometry could be useful for studying the early stages of sol-gel transitions.

These initial experiments indicate that important quantitative measurements, which can be used to obtain the gelation time by a non-intrusive method can be made. Furthemore, the spatial distribution of the gel primary particles and the influence of the thickness of the sample on the gelation process can also be observed.

In future experiments an interferometric analysis of thin sol-gel layers, such as those deposited as protective coatings on metal substrates, will be tried. In addition a new experimental set up to measure surface and thermal effects on the gelation processes, will be constructed.

REFERENCES

1. Ph. Colomban, Ceramic International   15 (1989) pp 23-50

2. C.J. Brinker and G.W. Scherer "Sol-Gel Science. The Physics and Chemistry of Sol-Gel Processing" Academic Press Inc.(1991) pp 160-216

3. N. Bauer et al. Techniques of Organic Chemistry, vol.I: Physical Methods of Organic Chemistry II. ed. A. Weissberger ,N.Y.; Interscience (1985) p.1139

4. C. Sainz, J. Calatroni and G. Tribillon; Meas.Sci. Tech. 1(1990) pp356-361

5. G. Tribillon, J. Calatroni and P. Sandoz; Recent Advances in Industrial Optical Inspection. SPIE. July 1990, San Diego, Cal. USA

THE STABILITY OF CERMET COATINGS AND THEIR DEPENDENCE ON THE DISTRIBUTION OF ELEMENTS

J.Lira, A.R.Di Giampaolo, I.C.Grigorescu, H.Ruiz and A.Sanz
Surface and Interface Engineering Research Group,
Material Science Department Simon Bolivar University
Caracas 1080, Venezuela

Abstract

The wear and corrosion of carbon steel parts can be reduced by coating them with an appropriate alloy. It would be even more desirable to design a cermet with toughness imparted by the metal matrix and hardness or strength by the ceramic reinforcement thus more closely approximating the industrial requirement for the sintered or coated parts. Investigation of the electrochemical and tribological behavior of commercial metal matrix composites and cermets has drawn attention to the chemical and mechanical behavior and characteristics of metal-ceramic interfaces.

This paper presents some adaptations and optimizations of techniques for fabricating and characterizing sintered composites of Ni-B-Si-VC. The influence of the distribution of the elements in the interface and in the bulk, and of the morphology of the precursor particles on the mechanical and electrochemical properties of these cermets, plus the potential of their use as protective coatings are discussed.

Using conventional sintering techniques and sintering + cold deformation + heat treatment, along with analysis by SEM, EDX, and optical microscopy, two major phases were distinguished in the matrix alloy. One, rich in Ni, resembled the particles of the original alloy powder. The other phase, where the carbides in the composite were embedded, was rich in B. Mechanical treatment tended to improve particle to particle cohesion, and thus the toughness of the metal. As a result of these findings, a procedure to improve thermal sprayed coatings is suggested.

INDEX OF CONTRIBUTORS